MATHLETICS

John D. Barrow is Professor of Mathematical Sciences and Director of Millennium Mathematics Project at Cambridge University, Fellow of Clare Hall, Cambridge, a Fellow of the Royal Society, and the current Gresham Professor of Geometry at Gresham College, London. His previous books include *The Book of Nothing*, *The Constants of Nature*, *The Infinite Book*, *Cosmic Imagery*, *100 Essential Things You Didn't Know You Didn't Know* and, most recently, *The Book of Universes*.

ALSO BY JOHN D. BARROW

Theories of Everything

The Left Hand of Creation
(with Joseph Silk)

L'Homme et le Cosmos
(with Frank J. Tipler)

The Anthropic Cosmological Principle
(with Frank J. Tipler)

The World within the World

The Artful Universe

Pi in the Sky

Perché il mondo è matematico?

Impossibility

The Origin of the Universe

Between Inner Space and Outer Space

The Universe that Discovered Itself

The Book of Nothing

The Constants of Nature:
From Alpha to Omega

The Infinite Book:
A Short Guide to the Boundless,
Timeless and Endless

New Theories of Everything

Cosmic Imagery:
Key Images in the History of Science

100 Essential Things You Didn't Know You Didn't Know

The Book of Universes

JOHN D. BARROW

Mathletics

A Scientist Explains 100 Amazing Things About Sport

VINTAGE BOOKS
London

Published by Vintage 2013
2 4 6 8 10 9 7 5 3 1

First published in Great Britain in 2012 by
The Bodley Head

Vintage
Random House, 20 Vauxhall Bridge Road,
London SW1V 2SA

www.vintage-books.co.uk

Addresses for companies within The Random House Group Limited can be found at: www.randomhouse.co.uk/offices.htm

The Random House Group Limited Reg. No. 954009

A CIP catalogue record for this book
is available from the British Library

ISBN 9780099584230

The Random House Group Limited supports the Forest Stewardship Council® (FSC®), the leading international forest-certification organisation. Our books carrying the FSC label are printed on FSC®-certified paper. FSC is the only forest-certification scheme supported by the leading environmental organisations, including Greenpeace. Our paper procurement policy can be found at www.randomhouse.co.uk/environment

Typeset in Dante MT by Palimpsest Book Production Limited,
Falkirk, Stirlingshire
Printed and bound by CPI Group (UK) Ltd, Croydon, CR0 4YY

'Heck, gold medals, what can you do with them'

Eric Heiden

TO MAHLER

who can already run
and soon will count

Contents

100 Essential Things You Didn't Know You Didn't Know About Sport

Preface

In this Olympic year I have taken the opportunity to demonstrate some of the unexpected ways in which simple mathematics and science can shed light on what is going on in a wide range of sporting activities. The following chapters will look into the science behind aspects of human movement, systems of scoring, record breaking, paralympic competition, strength events, drug testing, diving, riding, running, jumping and throwing. If you are a coach or a competitor you may get a glimpse of how a mathematical perspective can enrich your understanding of your event. If you are a spectator or commentator then I hope that you will develop a deeper understanding of what is going on in the pool, gymnasium or stadium, on the track or on the road. If you are an educator you will find examples to enliven the teaching of many aspects of science and mathematics, and to broaden the horizons of those who thought that mathematics and sport were no more than a timetable clash. And if you are a mathematician you will be pleased to discover how essential your expertise is to yet another area of human activity. The collection of examples you are about to read covers a great many sports and tries to pick topics that have not been discussed extensively before. Occasionally, there is a little bit of Olympic history for perspective, but it is balanced by chapters about several non-Olympic sports as well, and if you wish to delve deeper with your reading or push a calculation further there are notes to show you where to begin.

I would like to thank Katherine Ailes, David Alciatore, Philip

Aston, Bill Atkinson, Henry Baker, Melissa Bray, James Cranch, John Eckersley, Marianne Freiberger, Franz Fuss, John Haigh, Jörg Hensgen, Steve Hewson, Sean Lip, Clement McCalla, Justin Mullins, Kay Peddle, Stephen Ryan, Jeffrey Shallit, Owen Smith, David Spiegelhalter, Ian Stewart, Will Sulkin, Rachel Thomas, Roger Walker, Peter Weyand and Peng Zhao for the help, discussions and useful communications that helped this book come into being. A few of the topics covered here have been presented in lectures at Gresham College in London and as part of the Millennium Mathematics Project's activity for the London 2012 Olympics. I am most grateful to these audiences for their interest, questions and input. I must also thank family members, Elizabeth, David, Roger and Louise for their enthusiasm – although it turned to disbelief when they realised that this book wasn't going to help them get any Olympic tickets.

John D. Barrow, Cambridge 2013.

1

How Usain Bolt Could Break His World Record With No Extra Effort

Usain Bolt is the best human sprinter there has ever been. Yet, few would have guessed that he would run so fast over 100m after he started out running 400m and 200m races when in his mid teens. His coach decided to shift him down to running 100m one season so as to improve his basic sprinting speed. No one expected him to shine there. Surely he is too big to be a 100m sprinter? How wrong they were. Instead of shaving the occasional hundredth of a second off the world record, he took big chunks out of it, first reducing Asafa Powell's time of 9.74s down to 9.72 in New York in May 2008, and then down to 9.69 (actually 9.683) at the Beijing Olympics later that year, before dramatically reducing it again to 9.58 (actually 9.578) at the 2009 Berlin World Championships. His progression in the 200m was even more astounding: reducing Michael Johnson's 1996 record of 19.32s to 19.30 (actually 19.296) in Beijing and then to 19.19 in Berlin. These jumps are so big that people have started to calculate what Bolt's maximum possible speed might be. Unfortunately, all the commentators have missed the two key factors that would permit Bolt to run significantly faster without any extra effort or improvement in physical conditioning. 'How could that be?' I hear you ask.

The recorded time of a 100m sprinter is the sum of two parts:

the reaction time to the starter's gun and the subsequent running time over the 100m distance. An athlete is judged to have false-started if he reacts by applying foot pressure to the starting blocks within 0.10s of the start gun firing. Remarkably, Bolt has one of the longest reaction times of leading sprinters – he was the second slowest of all the finalists to react in Beijing and third slowest in Berlin when he ran 9.58. Allowing for all this, Bolt's average running speed in Beijing was 10.50m/s and in Berlin (where he reacted faster) it was 10.60m/s. Bolt is already running faster than the ultimate maximum speed of 10.55m/s that a team of Stanford human biologists recently predicted for him.[1]

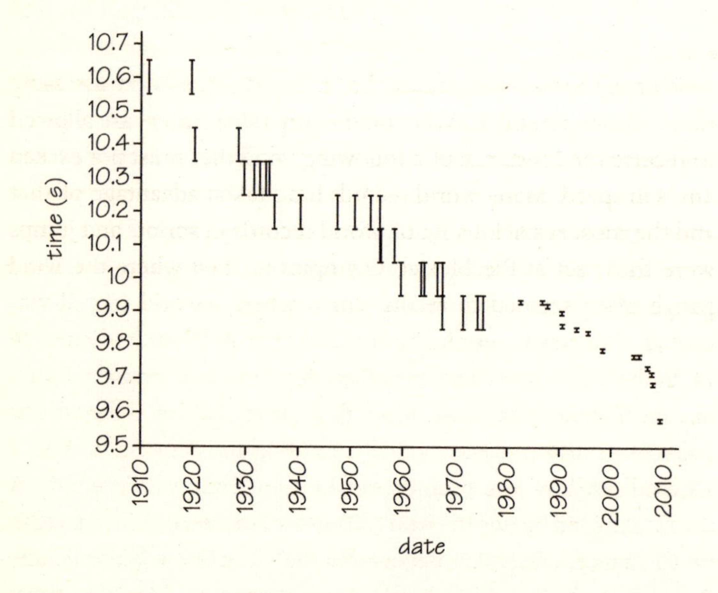

Bolt	0.146 + 9.434 = 9.58	Thompson	0.119 + 9.811 = 9.93
Gay	0.144 + 9.566 = 9.71	Chambers	0.123 + 9.877 = 10.00
Powell	0.134 + 9.706 = 9.84	Burns	0.165 + 9.835 = 10.00
Bailey	0.129 + 9.801 = 9.93	Patton	0.149 +10.191 = 10.34

In the Beijing Olympic final, where Bolt's reaction time was 0.165s for his 9.69 run, the other seven finalists reacted in 0.133, 0.134, 0.142, 0.145, 0.147, 0.165 and 0.169s.

From these stats it is clear what Bolt's weakest point is: he has a very slow reaction to the gun. This is not quite the same as having a slow start. A very tall athlete, with longer limbs and larger inertia, has got more moving to do in order to rise upright from the starting blocks.[2] If Bolt could get his reaction time down to 0.13, which is very good but not exceptional, then he would reduce his 9.58 record run to 9.56. If he could get it down to an outstanding 0.12 he is looking at 9.55 and if he responded as quickly as the rules allow, with 0.1, then 9.53 is the result. And he hasn't had to run any faster!

This is the first key factor that has been missed in assessing Bolt's future potential. What are the others? Sprinters are allowed to receive the assistance of a following wind that must not exceed 2m/s in speed. Many world records have taken advantage of that and the most suspicious set of world records in sprints and jumps were those set at the Mexico Olympics in 1968 where the wind gauge often seemed to record 2m/s when a world record was broken. But this is certainly not the case in Bolt's record runs. In Berlin his 9.58s time benefited from only a modest 0.9m/s tailwind and in Beijing there was nil wind, so he has a lot more still to gain from advantageous wind conditions. Many years ago, I worked out how the best 100m times are changed by wind.[3] A 2m/s tailwind is worth about 0.11s compared to a nil-wind performance, and a 0.9m/s tailwind 0.06s, at a low-altitude site. So, with the best possible legal wind assistance and reaction time, Bolt's Berlin time is down from 9.53s to 9.47s and his Beijing time becomes 9.51s. And finally, if he were to run at a high-altitude site like Mexico City, then he could go faster still and effortlessly shave off another 0.07s.[4] So he could improve his 100m time to an amazing 9.4s without needing to run any faster.[5]

2

All-rounders

Humans are often compared rather unfavourably with the champions of the animal kingdom: cheetahs sprinting faster than the motorway speed limit, ants carrying many times their body weight, squirrels and monkeys performing fantastic feats of aerial gymnastics, seals that swim at superhuman speeds, and birds of prey that can pluck pigeons out of the air without the need for guns. It is easy to feel inadequate. But really we shouldn't. All these stars of the animal kingdom are really nowhere near as impressive athletes as humans. They are very good at very special things and evolution has honed their ability to dominate their competitors in a very particular niche. We are quite different. We can swim for miles, run a marathon, run 100m in less than ten seconds, turn a somersault, ride a bike or a horse, high jump over eight feet, shoot accurately with rifles and bows, throw small objects nearly a hundred metres, ride a bicycle for hundreds of kilometres, row a boat, and lift much more than our body weight over our heads. Our range of physical prowess is exceptional. It's easy to forget that no other living creature can match us for the diversity of our physical abilities. We are the greatest multi-eventers on earth.

3

The Archers

Olympic archery is a dramatic participation sport but it is not so easy to see what is happening without a good pair of binoculars or big video monitors to replay the shots. The archers shoot seventy-two arrows at a circular target 70m away. The target is 122cm in diameter and divided into ten concentric rings, each of which is 6.1cm wide.

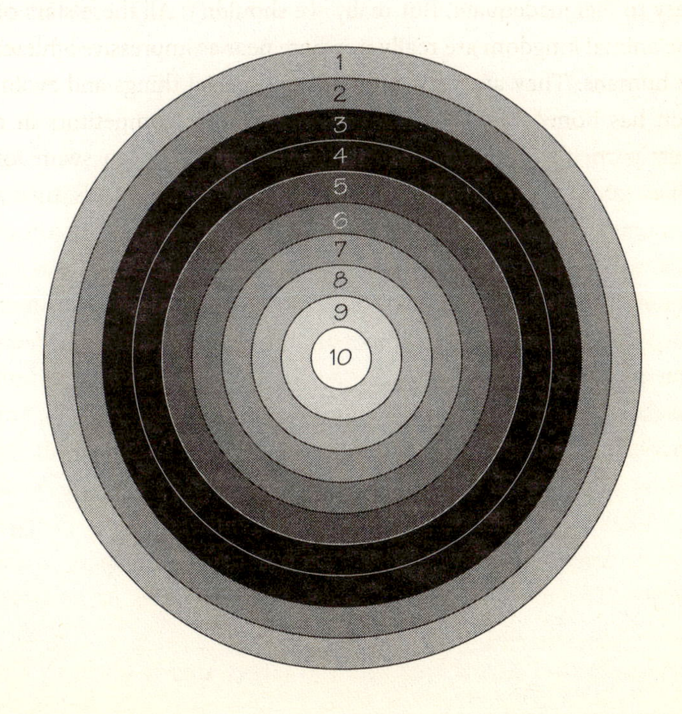

The two inner rings are gold and arrows landing there score 10 and 9 points. Going outwards the next two are red and score 8 and 7 points; the next two are blue and score 6 and 5; the next two are black and score 4 and 3; the last two are white and score 2 and 1. If you hit the target further out than this (or miss it completely) you score zero. These coloured circles are printed on a 125cm × 125cm square of paper that is backed by a protective layer to stop the arrows from penetrating through it.

The world's best archer is the South Korean woman Park Sung-Hyun. She scored a total of 682 points from seventy-two arrows to win individual and team gold medals at the 2004 Athens Olympics.[1] If she only scored 10s and 9s with all her arrows we can work out how she would have achieved that score. If T arrows scored 10 and the other 72–T arrows scored 9 then we know that $10T + 9(72-T) = 682$ and so $T = 34$ gives the number of 10s scored. The number of 9s would have been $72-34 = 38$. If she only scored 10s, 9s and 8s you might like to show that she must have scored thirty-five 10s, thirty-six 9s and one 8.

The difficulty of getting a particular score with one arrow is determined by the area of the annular ring that you have to hit to obtain it. The outer radii (in centimetres) of each of the ten circular rings are 6.1, 12.2, 18.3, 24.4, 30.5, 36.6, 42.7, 48.8, 54.9 and 61. Since the area of a circle is just π (= 3.14) times the square of its radius we can work out the area of each annular ring by subtracting the area of its inner bounding circle from the area of its outer bounding circle. So, for example, the area of the ring in which arrows score 9 points is $\pi\,(12.2^2 - 6.1^2) = \pi \times 6.1 \times 18.3 = 350.7$. I won't work out the areas of all the target rings but the same principle gives them very easily. Now, the likelihood of your arrow gaining a particular score is given by the fraction of the target area occupied by that part of the target. The area of the whole circular target is $\pi \times 61^2 = 11689.9$sq cm and so the probability of scoring a 9 with a randomly shot arrow that hits the target area is given by the ratio of the area in which you score 9 to the total area and this is 350.7/11689.9 = 0.03, or 3%. If I do these sums for the relative areas of all the scoring rings I get the probabilities that randomly shot arrows will hit any one of them. There is a simple pattern. The probabilities rise by 2% per ring as you move outwards through the rings. The hardest to hit is the centre ring with a 1% (i.e. 0.01) chance for a random shot; the easiest is the outer ring with a 19% (i.e. 0.19) chance of scoring 1 point.

If we add up all these average contributions we get 3.85 as the score we are likely to get from shooting a single arrow randomly at the target. If we shoot seventy-two arrows randomly then the average score we will get will be seventy-two times this, or 277, to the nearest round number. As you might expect this is far, far less than the world record score of 682. A score of 277 is what you would achieve with a purely random shooting strategy with no skill at all (except to hit some part of the target).

In calculating this we assumed that a random archer always hits the circular target. Suppose that they are not even that accurate and end up hitting anywhere at random inside the 125cm × 125cm square on which the target is printed. Its area is 15,625sq cm and you score zero if you hit this square beyond the outer circle of radius 61cm. In this case, all the overall probabilities and scores are reduced by a factor equal to the ratio of the area of the outer circle divided by the square, which is 11689.9 ÷ 15625 = 0.75. Therefore the average score obtained by shooting seventy-two arrows at random within the bounding square falls to 207.4.

If you want to test your arithmetic then you can apply exactly the same principles to calculate what score would be obtained by a random darts player. You should find that the average score is 13 points per dart, giving a score of 39 for three darts.[2]

4

The Flaw of Averages

Averages are funny things. Ask the statistician who drowned in a lake of average depth equal to 3cm. Yet, they are so familiar and seemingly so straightforward that we trust them completely. But should we? Let's imagine two cricketers. We'll call them, purely hypothetically, Anderson and Warne. They are playing in a crucial Test match which will decide the outcome of the series. The sponsors have put up big cash prizes for the best bowling and batting performances in the match. Anderson and Warne don't care about batting performances – except in the sense that they want to make sure there aren't any good ones at all on the opposing side – and are going all out to win the big bowling prize.

In the first innings Anderson gets some early wickets but is taken off after a long spell of very economical bowling and ends up with figures of 3 wickets for 17 runs, an average of 5.67. Anderson's side then have to bat and Warne is on top form, taking a succession of wickets for final figures of 7 for 40, an average of 5.71 runs per wicket taken. Anderson therefore has the better (i.e. lower) bowling average in the first innings, 5.67 to 5.71.

In the second innings Anderson is expensive at first, but then proves to be unplayable for the lower-order batsmen, taking 7 wickets for 110 runs, an average of 15.71. Warne then bowls at Anderson's team during the last innings of the match. He is not as successful as in the first innings but still takes 3 wickets for 48 runs, for an average of 16. So, Anderson has the better average bowling performance in the second innings as well, this time by 15.71 to 16.

Bowler	First Innings figures	First Innings average	Second Innings figures	Second Innings average	Combined figures	Overall average
Anderson	3 for 17	5.67	7 for 110	15.71	10 for 127	12.7
Warne	7 for 40	5.71	3 for 48	16	10 for 88	8.8

Who should win the bowling man-of-the-match prize for the best figures? Anderson had the better average in the first innings and the better average in the second innings. Surely, there is only one winner. But the sponsor takes a different view and looks at the overall match figures. Over the two innings Anderson took 10 wickets for 127 runs for an average of 12.7 runs per wicket. Warne, on the other hand, took 10 wickets for 88 runs and an average of 8.8. Warne clearly has the better average and wins the bowling award despite Anderson having a superior average in the first innings and in the second innings!

5

Going Round the Bend

Have you ever wondered whether it's best to have an inside or an outside lane in track races like the 200m where you have to sprint around the bend? Athletes have strong preferences. Tall runners find it harder to negotiate the tighter curve of the inside lane than that of the gentle outer lanes. The situation is even more extreme when sprinters race indoors where the track is only 200m around, so the bends are far tighter and the lanes are reduced in width from 1.22m to 1m. This was such a severe restriction that it became common for the athlete who drew the inside lane for the final (by being the slowest qualifier on times) to scratch from the final in indoor championships because there was so little chance of winning from the inside and a considerable risk of injury. As a result, this event has largely disappeared from the indoor championship roster.

But what about the outdoor situation where the curve is not so extreme? Most athletes don't like to be right on the outside because you can't see anyone (unless they pass you) for the first half of the race and you can't run 'off' their pace. On the inside you have a metal kerb marking the inside of your lane and you tend not to get as close to it as you would to the simple white painted line that marks the inside of the other lanes. Generally, the fastest qualifiers from the previous round are placed in the centre two or three lanes – a clear signal that they might be advantageous. A runner's physique is a factor too. If you are tall and long-legged you will have a harder time in the inner lanes and may have to chop your stride or run towards the outside of your lane to run freely. Potentially even more significant is the wind. If the wind is blowing at right angles to the finishing straight, into the faces of the runners when they run around the bend, then you will want to be in the outside lane so that you will be starting some way around the bend and will not have to run directly into the wind for so long – unlike those runners on the inside.

Finally, it is easy to show that you need to work harder if you run in the inside lanes. The two bends of an athletics track are semicircular. The radius of the circle traced by the inner line of the inside lane is 36.5m and each lane is 1.22m wide. So, the radius of the circle that you run in gets larger and the extra force that you have to exert to run in a circular path gets smaller and you actually run a smaller part of a circle as well. The radius of the circle traced by lane eight is $36.5 + (7 \times 1.22) = 45.04$m. The force needed for a runner of mass m to run in a circular path of radius r at speed v is mv^2/r, so as r gets larger,[1] and the bend is less tight, the force needed to maintain a given speed v decreases. If two identical runners, one in lane one and the other in lane eight, exert the same force over the first 100m of a 200m race, then the runner in lane one will have achieved a speed that is about 0.9 of that achieved in lane eight and the runner in lane eight will take 0.9 of the time. This is a very large factor – worth a whole second off the time for the first half of the race if you are running a 20s time for 200m. In practice there isn't such a large systematic advantage to running in the outside lanes and the runner only has to supply a fraction of the full circular motion force to sprint around the curve.[2]

If this simple model were complete then all 200m runners would run their best times from the outside lane. In practice most records are set from lanes three and four. Even this fact is slightly biased because the fastest qualifiers for the finals of big championships will have been put in those lanes. Presumably, the psychological and tactical advantages of being able to see your opponents and judge your speed against them from an inside lane helps outweigh the mechanical advantage of running around a gentler curve.

A good final comparison to make which illustrates the effect of the curve on 200m runs is to compare the world records run on a straight track with those around a curve. Straight 200m tracks are very rare now. There used to be one at the old Oxford University track at Iffley Road (where Roger Bannister ran the first sub-four-minute mile in 1954) that was still there when I began as a student in 1974 but had been removed by the time I graduated in 1977. When Tommie Smith set his world 200m record of 19.83s around a curve at altitude in the 1968 Mexico Olympics, he had already run a remarkable[3] 19.5s on a straight cinder track in San Jose in 1966. This latter record was only beaten by Tyson Gay, who ran 19.41s at the Birmingham City Games in 2010, watched by a 65-year-old Smith. Gay's fastest time around a curve is 19.58s. These time differences show the considerable slowing that is created by negotiating the curve. You might be lucky and have the wind behind you all the way in a straight 200m, but nonetheless runners find it strange to sprint such a long way without the reference points of the curve and other runners to dictate where they are and how they should apportion their effort.

6

A Question of Balance

If there is one attribute that is invaluable in just about any sport, it is balance. Whether you are a gymnast on the beam, a high-board diver, a spinning hammer thrower, a rugby forward snaking through the opposition's defence, a wrestler, a judoka trying to throw an opponent or a fencer lunging forwards, it is all about balance. Try a little experiment to see how well balanced you are and get a feeling for the muscle control behind it. Just stand completely still with one foot immediately in front of the other, so that the heel of your left foot touches the toe of your right foot. You can shift your weight so that it is mainly over the front foot or the back foot but keep your hands by your sides. You will probably find that standing completely still in a relaxed way is surprisingly difficult and your calf muscles are being tensed this way and that all the time. If you spread your arms out sideways you will find it much easier to balance. But now try leaning to one side. You won't lean very far before you lose your balance completely. Now, if you move your feet apart, in a normal standing position so they are not one behind the other in a straight line, then you will find it easier still, even with your arms by your sides – this, after all, is probably your usual stance. Lastly, go back to that difficult position with one foot directly in front of the other, but slowly crouch down low. You will find that balancing gets easier as you get nearer to the ground.

These little exercises reveal some simple principles for maintaining a good balance:

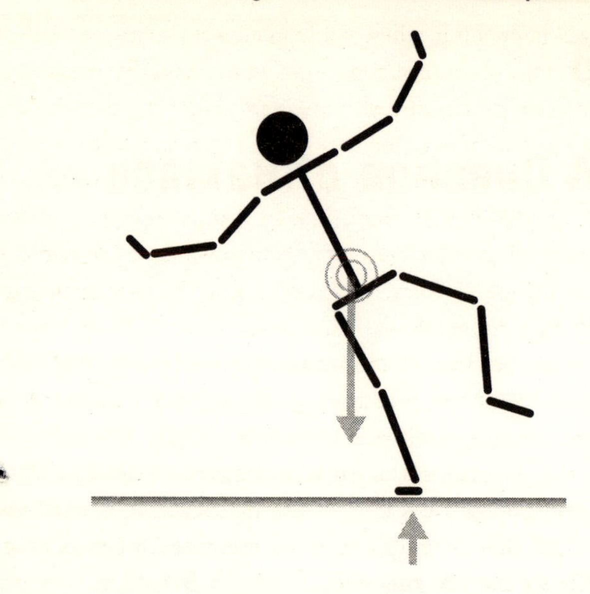

Make sure that the vertical line through your body's centre of gravity doesn't fall outside the base of support created by your feet. Once it does, you will fall away from equilibrium. You can experiment for yourself to see how far you can lean sideways, while keeping the body straight, before you start falling. The high-board diver will often use this instability in order to initiate his dive, leaning forward until his movement is taken over by gravity.

Broaden your base of support as much as possible. This makes it harder for your centre of gravity to fall outside your base. If you can stand on two feet, rather than on one, this will always help.

Keep your centre of gravity as low as possible. This is why you often see female gymnasts on the beam going into a low crouch position during a swing, perhaps with only one foot on the beam and one leg dangling below the beam – this lowers the centre of gravity even more. Sit astride the beam and you will see that balance is easy – your centre of gravity can't get much lower.

Spread your weight as far from your centre as possible. This is what was happening when you spread your arms out sideways. This is changing the distribution of your mass. By moving more of it far from your centre you are increasing your inertia, or your tendency not to move. Increasing your inertia in this way won't stop you wobbling but, crucially, it will make you wobble more slowly.[1] This gives you more time to take corrective action, shift your centre of gravity sideways or downwards, as required. This is why tightrope walkers carry long poles: they are ensuring that they wobble more slowly and have more time to correct a dangerous imbalance. Without that helpful pole, the man walking between skyscrapers on a high wire would surely fall to his death once he started to wobble in the breeze.

Watch wrestling and judo, where competitors are constantly trying to make their opponents lose their balance in subtle ways, or by using their strength to force them to violate one of the principles we have highlighted.

7

Anyone for Baseball, Tennis or Cricket?

A lot of people spend a lot of time hitting or chasing small spherical projectiles while dressed in unusual items of clothing. Games like baseball, tennis and cricket involve someone receiving one of these projectiles at very high speed. They have a split second to respond, either by getting out of the way, or hitting the projectile back as skilfully as they can. Which of these three sports requires the quickest reactions?

In each case the ball is different in size and can be launched by the pitcher, server or bowler at different speeds. Baseball has the simplifying feature that the ball only flies through the air, whereas in cricket and tennis it will hit the ground and rebound unpredictably because of its spin. In all three cases, the ball can swerve in the air to deceive the receiver in many ways. Let's ignore these extra degrees of difficulty and just focus on how quickly the receiver has to react to the incoming ball in each of these three games.

First, take cricket: a cricket pitch is 22 yards (= 20.12m) long.[1] The fastest bowlers achieve speeds exceeding 100mph, which is about 45m/s. The bowler will generally take a lengthy approach run in order to build up speed but the ball must be released with a straight arm or a 'no-ball' will be called for 'throwing'. If the batsman stands 1m in front of the wicket then he has 19.12/45 = 0.42s before the ball arrives at his bat.

By contrast, the baseball pitcher takes no run-up. He winds himself up on the spot from one of two allowed positions, the 'windup' or the 'stretch'. The counterpart of the fast bowler in cricket is the power pitcher. He relies on speed of delivery to outwit the hitter, who is 18.44m away from the pitcher when the ball is launched towards him. The quickest fastball that the best pitchers can deliver moves at about 100mph (= 45m/s). Unlike in cricket, bent arm pitching is allowed. The hitter's reaction time is therefore just 18.44/45 = 0.41s – a bit less than the cricket batsman gets.

What about the tennis player? Over time, racket technology has led to faster and faster serves to such an extent that top-class tennis is in danger of becoming dominated by service aces with few rallies. The record books list the fastest recorded serve by a man as 163.6mph (73m/s) by Bill Tilden in 1931. How this was measured I don't know. More reliable might be the value of 149mph (67m/s) attributed to Greg Rusedski at Indian Wells in 1998. The fastest recorded by a woman is 128mph (58m/s) by Venus Williams in 1998. The tennis court has a length of 78ft and for singles a width of 27ft. If the server and the receiver are located in opposite corners at the edges of the court then the distance travelled by the ball (neglecting differences in height above ground) is given by the length of the diagonal of a right-angled triangle with sides 78 and 27. Using Pythagoras' theorem, this is the square root of 6,084 + 729, which is 82.54ft, or 25.16m. If we assume that a top-flight serve is hit at 140mph (62.6m/s) and ignore any loss of speed when it bounces in the service court area then the receiver has about 25.16/62.6 = 0.40s to react.

The most interesting thing about our three rough calculations is not whether baseball players react one or two hundredths of a second quicker than tennis players or cricketers. Rather, it is the striking similarity between the required reaction times in the three different sports, to within a few hundredths of a second. Each has pushed the human response quite close to its limit.

8

Bayes Watch

Chance and probability play a major part in our lives. From court decisions about multiple cot deaths and DNA matches, to health and safety risks, you can't escape them. Determining likelihood is controversial and for the unwary beset by subtle pitfalls. Major miscarriages of justice have occurred because of ignorance by 'expert' witnesses in life and death judicial proceedings. The stakes are almost as high in the world of sport. Failing a drugs test can end your career and rob you of records, championships and multimillion-dollar commercial contracts. Thus testing athletes in a foolproof way is very important. There have been cases of incompetence by certified testing laboratories that have wrecked the careers of athletes and, in the case of Diane Modahl from 1994–8, brought about the collapse of the British Athletics Federation.

Baseball is an interesting case. Leading players are suspected of achieving their record-breaking feats by means of systematic steroid use. There is no random drug testing and disqualification system in place in US baseball, but anonymous testing has consistently revealed a worrying level of steroid use. Let's suppose that such a test reveals that 5% of players are steroid users and scientists tell us that the test is 95% accurate. What does that mean?

Suppose that 1,200 players are tested. Then we expect 60 (that's 5%) of them to be steroid users and the other 1,140 to be 'clean'. Of the 60 cheats, we expect 95%, that's 57 of them, to be correctly identified by the drug testers. But of the 1,140 clean competitors,

57 (that's 5% of 1,140) would be incorrectly recorded as drug-free by the testers.

These are sobering statistics. The testing of 1,200 players would result in 114 positives. Of these, 57 would be guilty of drug taking and 57 would not. So, if any player tests positive there is only a 50% chance that he or she has taken drugs.

What we have described here is an example of a very important piece of reasoning about conditional probabilities, first pointed out by the Reverend Thomas Bayes of Tunbridge Wells in 1763, in an article entitled 'Essay Towards Solving a Problem in the Doctrine of Chances'. What Bayes tells us is the relationship between the probability that an athlete is a drug taker given a failed test and the probability that a drug taker fails the test. Suppose we label by E the event that the drug test is positive, and by F the event that an athlete is taking drugs, then:

P(E) = probability that a drug test is positive
P(F) = probability that an athlete is a drug taker
P(E/F) = probability that a drug-taking athlete tests positive
P(F/E) = probability that athlete has taken drugs if their test is a positive

It is very important to recognise that P(E/F) and P(F/E) are different things. Prosecuting counsels in the courts are notorious for trying to bamboozle jurors into thinking they are the same, an error that is called 'the prosecutor's fallacy'.

What we want to know is P(F/E). In our example with the 1,200 players we know that P(F) = 0.05 and so the probability that an athlete is not using drugs is given by P(not F) = 0.95. The accuracy rate for the test is 95% and so P(E/F) = 0.95. We saw that 57 out of 1,200 (i.e. 4.75%) drug-free athletes tested positive and so P(E/notF) = 0.0475. What the Reverend Bayes showed is that all these quantities are related by a single formula:

$$P(F/E) = \{P(E/F) \times P(F)\}/\{P(E/F) \times P(F) + P(E/\text{not } F) \times P(\text{not } F)\}$$

In our example, this becomes:

$$P(F/E) = \{0.95 \times 0.05\}/\{0.95 \times 0.05 + 0.0475 \times 0.95\} = 0.513$$

So Bayes' formula shows that P(F/E) is completely different to P(E/F). In our example P(F/E) is unacceptably small and far better testing would be required to discriminate more accurately between drug takers and clean athletes.

9

Best of Three

Suppose that a football match takes place between the Reds and the Blues and the probability that the Reds score a goal is p and the probability that the Blues score a goal is 1–p. If an odd number of goals were scored what is the probability that the Reds won the game?

Well, if only one goal is scored then the probability that the Reds win the game is just p, the probability they score that single goal. But what if three goals are scored? The possible scoring sequences by the Reds (R) and Blues (B) and final results are as follows:

RRR	3-0	RBB	1-2
RRB	2-1	BRB	1-2
RBR	2-1	BBR	1-2
BRR	2-1	BBB	0-3

The probability of each of these eight results occurring is obtained by multiplying the probabilities for each goal occurring,[1] so for example, the probability of BBR is $(1-p) \times (1-p) \times p = (1-p)^2p$ and so on.

What is the probability of the Reds winning a three-goal game? It is simply the sum of the probabilities of the four ways in which they can win the game: RRR with probability p^3, plus that for the sequences RRB, RBR and BRR, each with probability $p^2(1-p)$. The total probability that the Reds win a three-goal game is therefore:

$$\mathbf{P}\text{ (Red win)} = p^3 + 3p^2(1-p) = p^2(3-2p)$$

If the Reds are the stronger team and much more likely to score, with $p = 2/3$, say, then their chance of winning the game is $\mathbf{P} = 20/27$, just slightly more than $2/3$ ($= 18/27$). If the two teams are evenly matched and $p = \frac{1}{2} + s$, where s is a very small quantity[2] then $p^2(3-2p)$ is approximately:

$$\mathbf{P}\text{ (Red win)} = \frac{1}{2} + 3s/2$$

If s is zero and the teams are equally likely to score then $\mathbf{P} = \frac{1}{2}$ too and they are equally likely to win the three-goal match. But if s is positive there will be a very small bias towards the Reds scoring a goal which gets amplified into a greater likelihood of $3s/2$ that they will win the match. We see that they are more likely to win when three goals are scored than if they are in a game where only one goal is scored. This doesn't mean that the Reds *will* win the match of course. Sometimes the weaker team does win, but the more games are played the greater the chance that the 'better' team will win in the long run.

10

High Jumping

There are two athletics events where you try to launch the body the greatest possible height above the ground: high jumping and pole-vaulting. This type of activity is not as simple as it sounds. Athletes must first use their strength and energy to launch their body weight into the air in a gravity-defying manner. If we think of a high jumper as a projectile of mass M launched vertically upwards at speed U then the height H that can be reached is given by the formula $U^2 = 2gH$, where g is the acceleration due to gravity. The energy of motion of the jumper at take-off is $\frac{1}{2} MU^2$ and this will be transformed into the potential energy MgH gained by the jumper at the maximum height H. Equating the two gives $U^2 = 2gH$.

The tricky point is the quantity H – what exactly is it? It is not the height that is cleared by the jumper. Rather, it is the height through which the jumper's centre of gravity is raised, which means it is a rather subtle thing, because it is possible for a high-jumper's body to pass over the bar even though his centre of gravity passes under it.

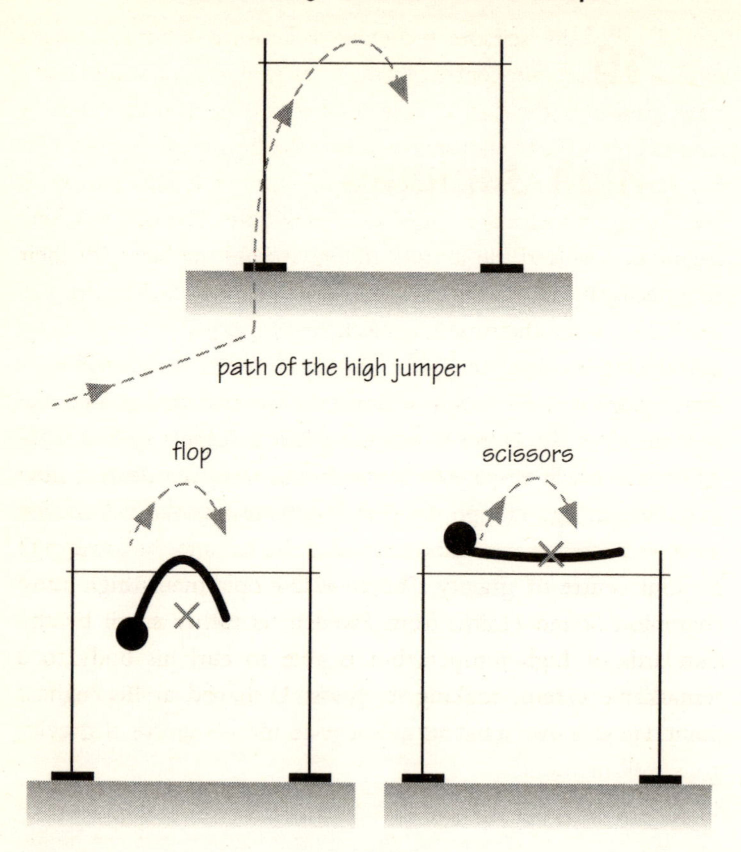

When an object has a curved shape, like an L, it is possible for its centre of gravity to lie outside of the body. This possibility allows a high jumper to control where his centre of gravity lies, and the trajectory it follows when he jumps. The high-jumper's aim is to get his body to pass cleanly over the bar whilst making his centre of gravity pass as far underneath the bar as possible. In this way he will make optimal use of his explosive take-off energy to increase H.

The simple high-jumping technique that you first learn at

school, called the 'scissors' technique, is far from optimal. In order to clear the bar your centre of gravity, as well as your whole body, must pass over the bar. In fact your centre of gravity probably crosses about thirty centimetres above the height of the bar. This is a very inefficient way to clear a high-jump bar and techniques used by top athletes are much more elaborate. The old 'straddle' technique involved the jumper rolling around the bar with their chest facing it and was the favoured method of world-class jumpers until 1968, when the American Dick Fosbury amazed everyone by introducing a completely new technique – the 'Fosbury Flop' – which employed a backwards flop over the bar. It won him the gold medal at the 1968 Olympics in Mexico City. Fosbury's technique was much easier to learn than the straddle and it is now used by every good high jumper. The more flexible you are the more you can curve your body around the bar and the lower will be your centre of gravity. The 2004 Olympic men's high-jump champion Stefan Holm, from Sweden, is rather small by the standards of high jumpers but is able to curl his body to a remarkable extent, making it almost U-shaped at his highest point. He sails over a bar set at 2m 40cm but his centre of gravity goes well below it.

11

Having the Right Birthday

Successful sportsmen and women are exceptional. Often the difference between success and failure in key competitions at any level is very small and anything that could give an advantage is important. Most top competitors are introduced to their particular sport while they are at school. They will have competed in school events and championships, and probably also joined a club outside of school, earning selection for county, regional or national teams in age-group competitions. In the UK, school age-group competition is usually based on the school year dates, starting on 1 September. In Europe, or for international competition, it is based on a calendar year, with a 1 January start date. Whatever the system employed, competitors in the same year group may differ in age by up to a year. When the competition age groups span two-year bands (15–17 or 17–19) this age gap doubles. For teenagers with varying growth rates and states of physical maturity, these age differences are very significant. As a result, children born in the first quarter of the school year (September to December) are on the average going to be bigger and stronger than those born in later quarters and therefore have a real advantage. They will tend to get into school teams or be selected for special coaching and are more likely to become successful competitors than their younger classmates. We would expect this bias to feed through into the proportions of teenagers who retain interest in their sport to become mature athletes. Several studies of the birthdates of successful sportsmen and women have indeed shown that first-quarter birthdays are favoured and clearly reveal the birthday bias.[1]

12

Air Time

Most people believe that if you want a projectile to go as far as possible then you should launch it (at least from ground level) at 45 degrees. This is a pretty good approximation to the truth if air resistance isn't a major factor. It is a useful rule of thumb to know if you are fielding near the cricket boundary or about to take a goal kick – both cases where achieving maximum range might be an important consideration. However, sometimes you want to invest in time rather than distance. When a kicker starts a rugby match from the centre spot he tries to loft the kick so that it stays in the air for a long time, so his players can arrive in numbers at the location of the unfortunate opponent who is waiting to catch it; likewise, a 'Garrryowen'[1] style up-and-under from open play allows lots of time for attackers to pressure an opposing defence. Footballers look to float free kicks and corners into the goalmouth area so that their team mates have more time to congregate in dangerous attacking areas.

When you kick a rugby ball at speed V at an angle θ to the ground it will fly along a parabolic path for a distance $R = V^2/g \times \sin(2\theta)$, where $g = 9.8\text{m/s}^2$ is the acceleration due to gravity, before returning to the ground. The maximum range is achieved when $\sin(2\theta)$ takes its maximum value of 1, which occurs when 2θ is 90 degrees and so θ when 45 degrees. Suppose that you specify that you just want the ball to end up in a particular place. This means fixing the value of the range, R.

We can now see that there are *two* ways of achieving this

because the sin of the angle A is the same as the sin of the angle (180 – A) degrees, so $\sin(2\theta) = \sin(180 - 2\theta)$, and the range R will be the same for both the launch angles θ and $90-\theta$ degrees. For example, a shallow launch angle of 15 degrees achieves the same range as a high trajectory of 75 degrees, although the time that the high ball is in the air will be longer if the launch speed is the same. The two trajectories are shown in the picture:[2]

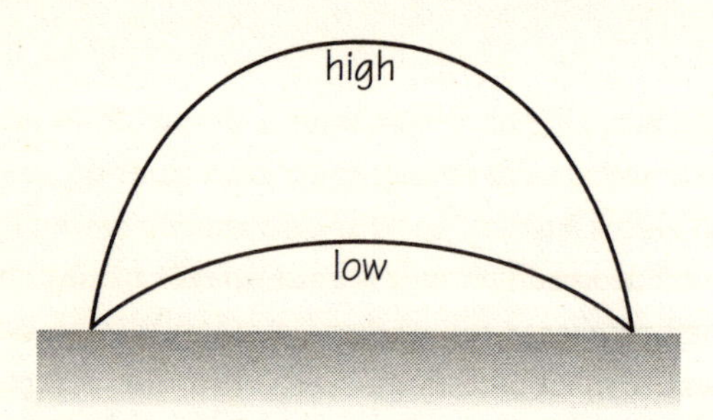

The time the ball takes to go a distance R is given by $R/(V\cos\theta)$. So, the ratio of the times of flight for the two kicks is t(high)/t(low) $= \cos(\theta)/\cos(90-\theta) = \cot(\theta)$ and the maximum vertical heights achieved by the two kicks is h(high)/h(low) $= \cot^2(\theta)$. We see that a high-trajectory ball launched at 75 degrees stays in the air 3.7 times longer and goes fourteen times higher than the shallow-trajectory ball launched at 15 degrees.

These considerations show how the simple geometry of projectile motions can allow players to 'play for time' in situations where you want extra time for your players to take up new positions on the pitch or outnumber defenders – or simply to dazzle defenders by having to receive a high ball coming out of the sun (or the floodlights).*

*We have ignored significant factors, notably wind, which will alter your strategy in important ways. In a crosswind the high trajectory will suffer significant lateral deviation. This will make it unpredictable for defenders and attackers and it is up to the attackers to make sure they reap the benefits!

13

Kayaking

Canoeing and kayaking have ancient origins in traditional communities and the word kayak is derived from the Inuit word *qajaq*. Actually, canoes are paddled on one side by kneeling crew members using single-bladed paddles, whilst kayak crews are seated and use double-bladed paddles, with alternate strokes on each side of the boat. Canoes are open but kayaks can be made completely watertight by the fitted clothing of the paddler to an extent that a complete 360-degree roll is possible by capsizing into the water and back out again without any water entering the kayak.

At the Olympics we see canoe and kayak races over a straight course of 500m or 1,000m with crews of one, two or four paddlers.[1] They are called C1, C2, K1, K2 etc. to denote the boat type and the number of paddlers. Unlike in rowing, there is never a cox; the paddlers have to steer their own course and they all face in the direction of motion.

If we look at the winning times at the Beijing Olympics in the 500m, the men's C1 was won in 1m 49.140s while the men's K1 was won in a significantly faster time, 1m 37.252s. The same trend is maintained for the K2 and C2 events. In fact, the female kayakers are a lot faster than the male canoeists over the same distance. Clearly, the extra stroke rate that is possible with the double blade, as well as the more streamlined profile of the kayak, gives a consistently quicker ride through the water and more efficient propulsion for the same amount of human power input.

But does having lots of paddlers help or hinder? The kayak

with two paddlers has twice as many 'engines' to power it but it has almost twice as much weight to drag through the water. Which is the dominant factor?

The power required to move the boat through the water is mainly to overcome the frictional drag on the hull created by the water and is equal to the drag force, D, times the speed, V, through the water. The drag force depends on the area of the hull in contact with the water, $A \propto L^2$, where L is the length of the boat, and the square of the speed through the water, so that:

$$\text{Power needed} = D \times V \propto L^2V^2 \times V \propto L^2V^3$$

Now, if there are N paddlers on board, the volume of the boat is proportional to L^3 which is proportional to N, the number of crew (you need a bigger boat to fit them in) so $L \propto N^{1/3}$, and the

$$\text{Power needed to overcome the drag} \propto L^2V^3 \propto N^{2/3}V^3$$

But the power supplied by the paddlers is proportional to their number:

$$\text{Power supplied by crew} \propto N$$

Since the power required to overcome the drag is supplied by the paddlers, we must have $V^3N^{2/3} \propto N$, and so we know how the speed of the boat should change with an increased number of paddlers*: $V \propto N^{1/9}$.

Thus the increased power supplied by the extra crew slightly outweighs the effect of the increased weight to be shifted through

*This trend is characteristic of an anaerobic exertion, which applies for distances less than about 400m. Over much longer distances the exertion becomes aerobic and the power supplied by a paddler will be proportional to their muscle mass, and hence to their bodyweight, and then the power to weight ratio and the speed will become independent of the total weight of the crew.

the water – but not by much. The gain from having extra crew increases very slowly with any increase in N. If the boats are assumed to proceed at a constant speed (which is not quite true, especially over the longer distances) then we would expect the finish times to be given by the length of the race divided by V, so the race times, T, should vary with N as $T \propto N^{-1/9}$.

Does this simple rule hold good? It predicts that if we divide the winning time in the two-man event by that in the one-man event, and the winning time in the four-man event by that in the two-man event, we should find that each of these ratios is approximately equal to the cube root of the cube root of 2, or $2^{1/9}$ = 1.08. For the men's 1,000m K events, we have:

T(one man)/T(two man) = 206.32/191.81 = 1.08
T(two man)/T(four man) = 191.81/175.71 = 1.09

For the women's 500m K events we have:

T(one woman)/T(two woman) = 110.67/101.31 = 1.09
T(two woman)/T(four woman) = 101.31/92.23 = 1.10

The agreement with the prediction of $2^{1/9}$ = 1.08 using our simple model is remarkably good. Despite all the idiosyncrasies of wind and weather, paddling style and kayak design, the dominant factors determining racing times are evidently just the simple power and drag factors we considered.

14

Do You Need a Cox?

If we are rowing rather than kayaking then the same principles hold as in the previous chapter. We saw that the increased power from extra crew in the kayak narrowly wins out over the extra weight to be pushed through the water and increased drag to be overcome. But it's a close-run thing. The speed of the boat only increases as the ⅑th power of the number of paddlers in the boat (speed $\propto N^{1/9}$) and the time taken over the race distance decreases in the same proportion.

Rowing differs from kayaking in one interesting respect: kayaks don't have coxes. It is clear that adding a cox contributes no extra power to the boat but increases the weight, size and drag, so a coxed four should go slower than a coxless four. The advantage of a cox is that the rowers don't have to worry about steering, and can avoid using energy to right the boat if it deviates from the shortest route to the finish.[1] They can focus entirely on rowing. A cox also plays an important role in encouraging the crew and dictating the stroke rate. Could these contributions outweigh the presence of a 'deadweight' in the boat, albeit usually a very light one?

If we look at the winning times for the men's coxed and coxless pairs and fours at the 1980 Moscow Olympics as examples, it is clear that the improved steering and encouragement doesn't outweigh the negative effects of a non-rowing passenger: the coxless times are always quicker than the coxed.

Number of rowers	Coxed	Coxless	Time ratio coxed / coxless
Pair: N = 2	422.5s	408.0s	1.04
Four: N = 4	374.5s	368.2s	1.02

If we repeat our previous look at the power and drag with one extra person contributing to the boat size and drag but not to the power generated, we find that the time to complete the course with N rowers plus one cox is:

$$T(\text{with cox}) \propto (N+1)^{2/9}/N^{1/3}$$

whereas the time without a cox is:

$$T(\text{coxless}) \propto N^{-1/9}$$

and their ratio gives:

$$T(\text{with cox})/T(\text{coxless}) = \{(N+1)/N\}^{2/9}$$

As expected, the quantity on the right-hand side is always greater than 1 (because N+1 is bigger than N) and so the time taken to race the same distance with a cox is predicted to be greater than without one. However, when we try to calculate how much slower the race time will be, we have to be careful. We have always been assuming our rowers (and kayakers) are all the same size. This is a pretty good approximation for the rowers but it isn't for the cox. You want your cox to be as small and lightweight as possible to reduce the extra load and drag as much as possible. It is a better approximation to assume that the cox is about half the weight of a rower. If we do that, then the cox contributes as N + ½ by weight to the total crew number rather than as N + 1 in our estimates for the overall racing time. As a result, a better estimate for the comparison racing times for the coxed and coxless boats is:

$$T(\text{with cox})/T(\text{coxless}) = \{(N + \tfrac{1}{2})/N\}^{2/9}$$

For the pairs (N = 2), this ratio is $(25/16)^{1/9} = 1.05$, while for the fours it is $(81/64)^{1/9} = 1.03$. These are very close to the ratios seen in the 1980 Olympic results shown in the table.[2]

Finally, one of the great mysteries in Olympic history revolves around a cox. In the 1900 Paris Games a Dutch coxed pair ditched their cox from the event because they thought he was too heavy. They picked a little French boy, about 10 years old, from the spectators to take his place. Despite their cox's inexperience they won the gold medal. Yet after the race, the child disappeared before anyone could find out who he was.

15

On the Cards

Collecting sets of cards was once all the rage. There were collections of wartime aircraft, animals, ships and sportsmen – since these collections all seemed to be aimed at boys – to be amassed from buying lots of packets of bubblegum, breakfast cereals or packets of tea. Of the sports cards, just like today's Panini 'stickers', the favoured game was football – in the US it was baseball – and I always had my suspicions about the assumption that all the players' cards were produced in equal numbers. Somehow everyone seemed to be trying to get the final 'Bobby Charlton' card that was needed to complete the set of 50. All the other cards could be acquired by swapping duplicates with your friends but everyone lacked this essential one – so you kept on buying bubblegum.

It is a relief to discover that even my own children engaged in similar acquisitive practices. The things collected might change but the basic idea was the same. So what has mathematics got to do with it? The interesting question is how many cards we should expect to have to buy in order to complete the set, if we assume that each of them is produced in equal numbers and so has an equal chance of being found in the next packet that you open.

The sports sets I came across each contained 50 cards. The first card I get will always be one I haven't got already but what about the second card? There is a 49/50 chance that I haven't already got it. Next time it will be a 48/50 chance and so on. After you have acquired 40 different cards there will be a 10/50 chance that the next one will be one you haven't already got. So, on the average.

you will have to buy another 50/10 = 5 cards to have a better-than-evens chance of getting another new one for your set. Therefore, the total number of cards you will need to buy on average to get the whole set of 50 will be the sum of 50 terms (50/50 + 50/49 + 50/48+ . . . +50/3 + 50/2 + 50/1), where the first term is the certain case of the first card you get and each successive term tells you how many extra cards you need to buy to get the second, third, and so on, missing members of the set of 50.

Taking out the common factor 50 in the numerators of each term, this is just 50(1 + ½ + 1/3 + . . . +1/50). The sum of terms in the brackets is the famous 'harmonic' series. When the number of terms in it becomes large (and 50 is large enough) it is well approximated by 0.58 + ln(50) where ln(50) = 3.9 is the natural logarithm of 50. So, we see that the number of cards that we need to buy on the average to complete our set is about:

$$\text{Cards needed} \approx 50 \times [0.58 + \ln(50)]$$

For my set of 50 sports cards the answer is 224.5 and I should expect to have had to buy on average about 225 cards to make up my set. Incidentally, our calculation shows how much harder it gets to complete the second half of the collection than the first half. The number of cards that you need to buy in order to collect 25 cards for half a set is (50/50) + (50/49) + (50/48) + . . . + (50/26), which is the difference between 50 times the harmonic series summed to 50 and summed to 26 terms, so:

$$\text{Cards needed for half a set} \approx 50 \times [\ln(50) + 0.58 - \ln(25) - 0.58]$$
$$= 50\ln(2) = 0.7 \times 50 = 35$$

Or just 35 to get 25 of my set of 50. This means that I am going to need to buy about 225 – 35 = 190 to get the second 25.

I wonder if the original manufacturers performed such

calculations. They should have, because they enable you to work out the maximum profit you could expect to gain in the long run from marketing a set of cards of a particular number. It is only likely to be a maximum possible profit because collectors will trade cards and be able to acquire new cards by swapping rather than buying new ones. What impact can friends make?

Suppose that you have F friends and you all pool cards in order to build up F + 1 sets so that you have one each. How many cards would you need to do this? On the average, for a 50-card set, and if you share cards, the answer approaches:

$$50 \times [\ln(50) + F \ln(\ln 50) + 0.58]$$

On the other hand, if you had each collected a set without swapping you would have needed about $(F + 1)50[\ln(50) + 0.58]$ cards to complete F + 1 separate sets and the number of card purchases saved would be 156F. Even with $F = 1$ this is a considerable economy.

16

Wheels on Fire

Every sport that makes use of equipment has a technical interest in understanding how to make it better – although in many cases the bigger imperative seems to be to get everyone involved to buy a new set of kit each season. Cycling is one of the most familiar challenges for engineers and we have seen the introduction of bodysuits, new handlebar designs and disc wheels in the quest to shave hundredths of a second off racing times in the velodrome.

An interesting question to pose is whether you get more advantage from reducing the weight of the wheels or that of the bike frame. In order to make the bike move, the cyclist has to provide the kinetic energy $\frac{1}{2}Mv^2$ needed to move the total mass of the bike frame plus rider plus the wheels, M, forwards at speed v. But the rider also needs to supply the rotational energy needed to make the wheels spin, which is $\frac{1}{2}Iw^2$, where w is the angular speed of the spinning wheels and $v = rw$ where r is the radius of a wheel. The quantity I is the inertia of the wheel (we will assume that both wheels are the same, for simplicity, and the wheels don't skid). It tells us how hard it is to move the wheel, and the inertia is larger when the mass gets spread further from the wheel's hub. The inertia is always proportional to the mass, m, of the wheel and the square of the wheel's radius, r, so $I = bmr^2$, where $b = 1$ if all the mass is in the outer ring (ignore the spokes – they are very light compared with the rest of the wheel) but $b = \frac{1}{2}$ if the wheel is a solid disk.

The total energy that the rider needs to supply[1] to move the bike frame plus rider and the two wheels forwards at speed v and rotate the two wheels is therefore:[2]

$$\text{Total energy} = \tfrac{1}{2}(2m + m_{frame})v^2 + 2 \times \tfrac{1}{2}Iw^2$$

Since $I = bmr^2$ and $v/r = w$, we have:

$$\text{Total energy} = \tfrac{1}{2}v^2\{m_{frame} + 2(1+b)m\}$$

So for the traditional ring style of wheel with $b = 1$, the drain on the rider's energy by the motion of the wheels is proportional to four times the mass of a wheel; for the disk wheel, with $b = \tfrac{1}{2}$, it is proportional to three times the mass of a wheel. It is interesting that the radius of the wheel cancels out of our formula and a smaller wheel is not more advantageous unless it also has a smaller mass. Clearly, if you are engineering new materials to reduce the weight of the bike then the same mass reduction on each wheel is three or four times as beneficial as the same mass reduction on the frame.

17

Points Scoring

The decathlon consists of ten track and field events spread over two days. It is the most physically demanding event for athletes. On day one, the 100m, long jump, shot-put, high jump and 400m are contested. On day two, the competitors face the 110m hurdles, discus, pole vault, javelin and, finally, the 1,500m. In order to combine the results of these very different events – times and distances – a points system has been developed. Each performance is awarded a pre-determined number of points according to a set of performance tables. These are added, event by event, and the winner is the athlete with the highest points total after ten events. The women's heptathlon works in exactly the same way but with three fewer events (100m hurdles, high jump, shot, 200m, long jump, javelin and 800m).

The most striking thing about the decathlon is that the tables giving the number of points awarded for different performances are rather free inventions. Someone designed them first back in 1912 and they have subsequently been updated now and then. Daley Thompson missed breaking the decathlon world record by one point when he won the Olympic Games in 1984, but a revision of the scoring tables the following year increased his score and he became the new world record holder retrospectively! The current world record is 9,026 points set by Roman Sebrle of the Czech Republic in 2001.* If you broke the world record in each of the individual events you would score about 12,500 points. The best performances ever

*Amazingly, there is also a record of 7,897 points held by Robert Zmelik for a decathlon that must be completed in less than sixty minutes!

achieved for each event during decathlon competitions sum to a total score of 10,485. The points tables were set up in 1912 so that (approximately) 1,000 points would be scored by the world record for each event at the time. But records move on and now, for example, Usain Bolt's world 100m record of 9.58s would score him 1,202 decathlon points, whereas the fastest 100m ever run in a decathlon is 'only' 10.22s for a points score of 1,042. The current world record that would score the highest of all is Jürgen Schult's discus record of 74.08m, which accumulates 1,383 points.

All of this suggests some important questions. What would happen if the points tables were changed? What events repay training investment with the greatest points payoff? And what sort of athlete is going to do best in the decathlon – a runner, a thrower or a jumper?

The setting of the points tables has evolved over a long period of time and pays attention to world records, the standards of the top-ranked athletes and historical decathlon performances. However, ultimately it is a human choice and if a different choice was made then different points would be received for the same athletic performances and the overall winners might change. The 2001 Internatial Association of Athletics Federations (IAAF) scoring tables[1] have the following simple mathematical structure.

The points awarded (decimals are rounded to the nearest whole number to avoid fractional points) in each track event – where you want to give higher points for shorter times (T) – are given by the formula:

$$\text{Track event points} = A \times (B - T)^C$$

where T is the time recorded by the athlete in a track event and A, B and C are numbers chosen for each event so as to calibrate the points awarded in an equitable way. The quantity B gives the cut-off time at and above which you will score zero points so T is always less than B. Similarly, for the jumps and throws – where you want to give more points for greater distances (D) – the points formula for each event is:

$$\text{Field event points} = A \times (D - B)^C$$

The numbers A, B and C are different for each of the ten events and are shown in this table. You score zero points for a distance equal to or less than B, or a time equal to or greater than B. The distances are all in metres and the times in seconds.

Event	A	B	C
100m	25.4347	18	1.81
Long jump	0.14354	220	1.4
Shot-put	51.39	1.5	1.05
High jump	0.8465	75	1.42
400m	1.53775	82	1.81
110m hurdles	5.74352	28.5	1.92
Discus	12.91	4	1.1
Pole vault	0.2797	100	1.35
Javelin	10.14	7	1.08
1,500m	0.03768	480	1.85

To get a feel for which events are 'easiest' to score in, take a look at this table which shows what you would have to do to score 900 points in each event for 9,000-point total.

Event	900pts
100m	10.83s
Long jump	7.36m
Shot-put	16.79m
High jump	2.1m
400m	48.19s
110m hurdles	14.59s
Discus	51.4m
Pole vault	4.96m
Javelin	70.67m
1,500m	247.42s (= 4m 7.4s)

There is an interesting pattern in the decathlon formulae. The power index C is approximately 1.8 for the running events (1.9 for the hurdles), close to 1.4 for the jumps and vault, and close to 1.1 for the throws. The fact that $C > 1$ indicates that the points system is a 'progressive' one, that it gets harder to score points as your performance gets better. This is realistic. We know that as you get more expert at your event it gets harder to make the same improvement but beginners can easily make large gains. A 'regressive' points system would have $C < 1$, while a 'neutral' one would have $C = 1$. The IAAF tables are extremely progressive for the running events, fairly progressive for the jumps and vault, but almost neutral for the throws.

In order to get a feel for how the total points scored is divided across events, the figure below shows the division between the ten events for the averages of the all-time top hundred men's decathlon performances:

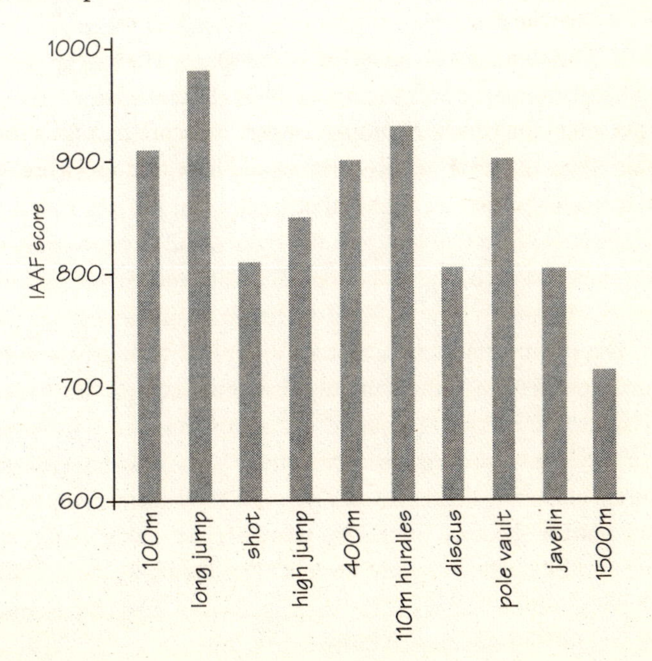

It is clear that there is a significant bias towards the long jump, hurdles and sprints (100m and 400m). Performances in these events are all highly correlated with flat-out sprinting speed. Conversely, the 1,500m and three throwing events are well behind the others. If you want to coach a successful decathlete then start with a big strong sprint hurdler and build up strength and technical ability for the throws later. No decathletes bother much with 1,500m preparation and just rely on general distance-running training.

Clearly, changes to the points-scoring formula would change the event. The existing formulae are based largely upon (recent) historical performance data of decathletes rather than on top performances by specialists in each event. This reinforces any bias in the current scoring tables because the top decathletes are where they are because of the current scoring tables. As an exercise, we could consider a simple change that is motivated by physics. In each event, whether it is sprinting, throwing or jumping – with the possible exception of the 1,500m – it is the kinetic energy generated by the athlete that counts. This depends on the square of his speed. The height cleared by the high jumper or pole-vaulter, or the horizontal distance reached by the long jumper, are all proportional to the square of his launch speed. Since the time achieved running at constant speed will be proportional to $(\text{distance}/\text{time})^2$ this implies that we pick C = 2 for all events. If we do that and pick the right A and B values then we get an interesting change in the top ten decathletes. Sebrle becomes number two with a new score of 9,318 whilst the present number two, Dvorak, overtakes him to take first place with a new world record score of 9,468. Other rankings change accordingly. The pattern of change is interesting. Picking C = 2 across all events is extremely progressive and greatly favours competitors with outstanding performances. However, it dramatically favours good throwers over the sprint hurdlers because of the big change in the value of C = 1.1 being applied to the throws at present. This illustrates the difficulty with points systems of any sort – there is always a subjective element that could have been chosen differently.

18

Diving

High-board diving is a sport that seems to have found itself paradoxically classified as a water sport. It is part of the schedule of swimming events at the Olympic Games even though it is a sport that takes place in the air and is closer to gymnastics or trampolining than it is to swimming or water polo.

There are two diving competitions: the high board from a fixed platform ten metres above the water, and the springboard which is three metres above the water. The high board is simpler in concept. Divers launch from a standing, running or handstand position in which their centre of gravity is about 1.2m above the board and 11.2m above the water. Ignoring all their somersaults and twists, for the moment, they will fall a distance s under gravity in a time given by $s = \frac{1}{2}gt^2$, where $g = 9.8m/s^2$ is the acceleration due to gravity. If the diver of height 1.8m enters the water head first with arms and hands outstretched then the diver's centre of gravity will be about 1.2m from their fingertips and they will have fallen approximately ten metres. Using this for s, we see that the time in the air was 1.4s. This is the time that the diver has to complete the series of somersaults and twists needed to impress the judges. And you will hit the water at a speed of 14m/s, or about 31mph. This will lead to a very painful collision with the water unless you streamline your entry to reduce the area of impact. A smooth, thin body profile on entry gives the water time to move out of the way and be displaced by the diver's body. Land inelegantly, or belly flop –

then ouch! It will be like hitting something much more rigid than water.

Suppose you want to complete three and a half somersaults in the air. This requires 3.5 revolutions in 1.4 seconds, or 2.5 rev/s. This is 150 revolutions per minute (rpm) and can be compared with a tracking speed of 200rpm at the outer edge of a disc playing on your CD player. Each revolution is 2π radians and so the angular velocity needed to complete the somersaults is $5\pi = 15.7$ radians/s.

The springboard is completely different. The board is three metres above the water but the diver launches upwards using the elastic energy stored in the board – but not exactly upwards or else there will be a nasty re-encounter with the springboard on the way down. Launching about 5 degrees away from the vertical at about 6m/s will produce a dive in which the centre of the diver follows a parabolic trajectory that peaks at a height of about six metres above the pool and gives the diver about 1.8s in the air to perform the necessary aerial gymnastics. Notice that this is longer than the 1.4s available to the high-board diver. Although the launch is seven metres closer to the pool, the decelerating upward trajectory buys lots of time for the diver and reveals how different these two events really are.

When divers complete their somersaults and twists they need to enter the water with a body position as vertical as possible, and with no rotation. This will create the smooth, splash-free entry that the judges are looking for. It requires meticulous timing and hundreds of hours of practice. The disturbance of the pool surface by sprays helps the divers judge where the water is located as they come out of their twists and turns. They reduce their rotation by doing the opposite of the trick that ice skaters perform so as to spin faster. In a spin, the product of the inertia and the angular speed of the spinning object is preserved, so by reducing your inertia you will spin faster. A skater's inertia is given by the product of their mass and their radius squared. By drawing their arms in as they spin the skaters can reduce their inertia by a factor of two and this will double the angular speed of their spin up to about 20 radians/s, or 3 revs/s.

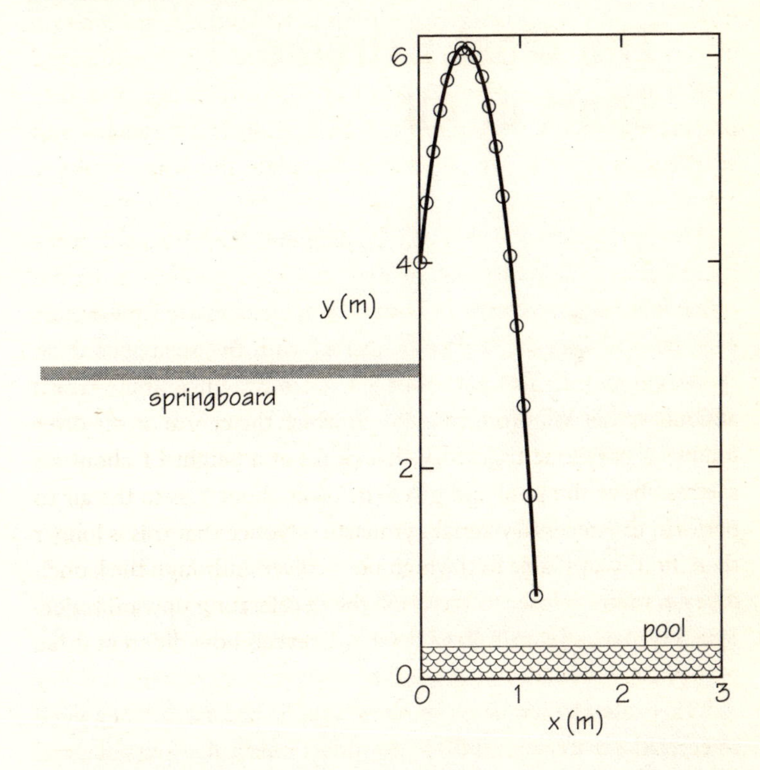

The diver spins faster in his somersaulting phases by tucking his body in tightly to reduce its radius and inertia. But at the end of the dive he straightens out his body completely so its length doubles, his inertia increases fourfold, and his angular speed is reduced by the same factor. Perform this manoeuvre precisely and you will have negligible rotation when you enter the water to complete your dive.

19

The Most Extreme Sport of All

What is the most extreme human activity? Astronaut, fighter pilot, Formula 1 racing driver, free-fall parachutist, Acapulco cliff diver, bobsleigh driver? The list could go on. But I think there is one candidate that trumps them all – drag-car racer. You won't see it at the Olympic Games and to watch it you will have to frequent disused airfields or salt flats in the middle of nowhere. Drag cars are like rockets with wheels attached, which race over a quarter of a mile (approximately 400m) in 4.5s from a stationary start. They accelerate faster than a NASA rocket launch and reach a top speed of over 330mph. If a Formula 1 racing car were to pass the start at top speed when the drag car starts from stationary, it would still be beaten to the finish by the dragster. The accelerations and decelerations experienced by the driver reach 6g – and greater during the deceleration phases after parachutes are released to slow the car down – and detached retinas are a serious problem for competitors. The noise levels are also dangerous for spectators, technicians and drivers. Good ear protection is absolutely essential.

Drag-car motion is interesting because, after a split second of rolling to pick up speed, the cars exhibit a motion under constant power which is supplied by the engine and the significant quantity of fuel on board. Power is force times velocity, or $mdV/dt \times V$, if we neglect the changing mass created by the fuel consumption.

V is the velocity of the car and m is its mass. If we set this equal to a constant power value P then[1] we find that the speed at time t is $V^2 = 2Pt$ and the distance travelled from the start, at $t = 0$, is $x = (8P/9)^{3/2}t^{3/2}$ if we assume the car starts from stationary. These formulae show that the speed achieved after going a distance x will be $V = (3Px/m)^{1/3}$. This is a rule of thumb in drag-car racing called 'Huntington's Rule', after the engineer Roger Huntington[2] – although we see that it is easy to prove it using simple mechanics. Drag-car mechanics use old-fashioned (non-metric but convenient) units of horsepower for P, mph for V, pounds for the car mass m. Huntington's Rule says that the speed in mph is equal to K times (power in horsepower/mass in pounds)$^{1/3}$. K is calculated to equal 270 in this simple description but measurements on cars find a value of about 225, not very different given how simple our model is.[3]

Our calculation shows that the drag car's speed increases like the square root of time elapsed and the cube root of the distance over its quarter-mile course.

20

Slip Slidin' Away

Slippery pitches are the bane of sports competitors. They produce unexpected results and unwanted injuries. In many sports the existence and understanding of the role of friction is crucial for competitors and for the companies manufacturing shoes, equipment and playing surfaces. Let's take a few examples. Discus throwers have to spin on a smooth cement surface; its smoothness must be right or throwers will skid or turn too slowly. Footballers and rugby players have to make sudden turns and accelerations on a grass surface that can behave very differently when wet or dry. Javelin throwers have to time their approach run so that they can stop suddenly after launching their spear. High jumpers need to place their take-off foot firmly on the take-off area surface without any possibility of slipping or serious injury will result.[1] Wrestlers can only hold their ground when pushing their opponents because of friction between their feet and the ring surface.

If you place your foot on the ground and try to move forwards by pushing backwards (this is usually called 'walking' or 'running') then friction is necessary if you are to move forwards. Friction arises because the points of contact between your footwear and the ground create some loose molecular bonds which need to be broken if your shoe is to part company with the ground again. Sometimes we see footballers suffering very serious knee ligament injuries because their studs cause a boot to stick in the turf when they drive their legs forward or at an angle. For a period, the new Wembley Stadium turf seemed particularly likely to cause such injuries and led to bitter complaints by rugby players and footballers playing there in cup finals and internationals. A higher level of surface friction will also tire you more quickly – roller hockey can seem a lot more tiring than ice hockey because it's harder to generate speed.

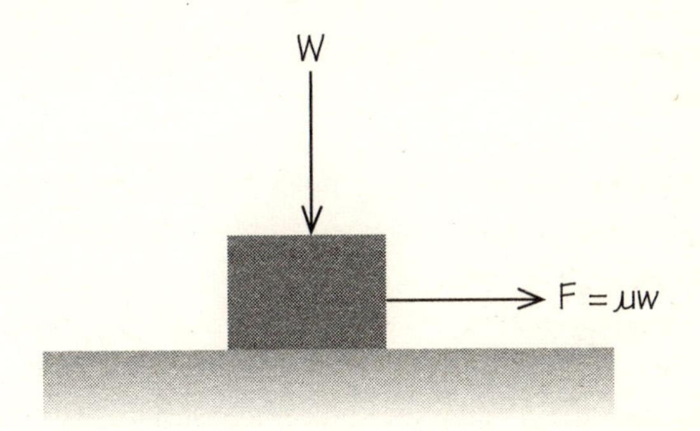

The frictional force opposing forward motion because of the molecular embrace between your boot and the ground is proportional to the vertical force you exert on the ground – your weight, W.[2] As the weight increases the materials in the two surfaces are pushed closer together and the molecular forces between them become stronger. The proportionality factor, μ, takes into account the intrinsic properties of the two surfaces: rubber on concrete is far less slippery than steel on ice. For most dry materials this factor lies between about 0.3 and 0.6. The smallest value for a familiar material is 0.04 for Teflon, of non-stick saucepan fame, and the largest values are actually between 1 and 2 for forms of silicone rubber. It's a good estimate to take the friction factor for leather or rubber sports shoes on a dry grass pitch to be about 0.3. When the pitch gets wet it might be reduced to about 0.2, and even less if it is icy.[3]

On the dry surface this means that if you start to run by exerting a force less than one third of your weight in a backwards direction then your foot will hold firm on the ground and you will be propelled forwards. However, if you apply a force greater than one third of your weight then this will exceed the frictional force between your boot and the pitch and you will slip. This places quite a limit on how fast you can run without slipping, and so in practice you will wear special sports shoes with studs, spikes or rippled soles to increase your grip on the surface and ensure that you can exert far greater backward force before you run the risk of slipping.

21

Gender Studies

There are only two sports at the Olympic Games where men and women compete against each other in open events. The first is equestrianism and the other, less well known, certain sailing events. Before 1984, the shooting events were open to men and women but separate competitions were introduced at the Games in Los Angeles that year. The first woman Olympic champion was the British tennis player, five-time Wimbledon champion Charlotte Cooper, who won the singles final on 11 July 1900 and then went on to win the mixed doubles with Reginald Doherty. Only eleven of the 1,077 competitors that year were women and only six women competed in St Louis in 1904.

The first equestrian events that entered the modern Games in 1900 were not ones that we see today. They included a horse high jump – two horses tied for first place with clearances of 1.9m, with one of the winners also riding the fourth-placed horse in this competition – and a horse long jump (won with 6.1m). Dressage, eventing and show jumping appeared in 1912 but only competitors who were commissioned officers in one of their country's military forces could compete in the dressage and had to wear their uniforms while competing. In fact, the Swedish gold medal team winners in 1948 were disqualified eight months after the Games because one of their team had not been an officer as they claimed. This odd restriction was abolished in 1952 when both women and civilians were allowed to compete in the dressage – although they had been allowed to compete in the show jumping ever since 1912 while military riders and horses were excluded from that event. The most demanding event, the three-day event, or horses' 'triathlon', allowed women to compete alongside the men since 1952; however, the first woman to compete, Helena du Pont, for the USA, would not be selected until 1964. Today, all the equestrian events are open to men and women and in the team competitions any mix of men and women is allowed. Equestrianism is also unique in being the only Olympic event where humans and animals compete together.

The equestrian events were totally dominated by male riders up until 1968, after which in the dressage there have been female gold, silver or bronze medallists at almost every Games. There are fewer in show jumping and they are all silver or bronze medallists. In the three-day event there have been female medallists at almost all Games since 1984. One expects no physical advantage for male competitors in the dressage but it is not obvious to an outsider that there are no differences between male and female competitors in the show jumping and three-day events, where significant strength and fitness is required as well as the obvious skill in controlling and managing the horse's enthusiasm, or reluctance, to jump fences. Yet, the fact that a 67-year-old Japanese competitor took part in Beijing (he had previously competed in the 1964 Games in Tokyo) shows that these events are not like others in the Olympics.

Sailing is another sport that is partly gender neutral. In the 2004 Games in Athens, there were four events for men only, four for women only, and three mixed (the Laser, 49er and Tornado classes), but no women won medals in the three mixed events. In Beijing, there was only one woman entrant and no medallist in the latter, which were now the Finn (won by Ben Ainslie), 49er and Tornado classes. The 2012 schedule shows that there will be six men's sailing classes and four women's (including one new one) with no open mixed classes for men and women any more. Maybe this will encourage more female participation.

Shooting seems like a natural candidate for a mixed-gender competition, and in fact all the events were open to both men and women until 1984, when specific women's events were introduced – although women could still compete with the men if there wasn't a dedicated women's competition in their event. Only two women won medals in these mixed events: Zhan Chan of China won gold in the Skeet[1] at the Barcelona Games of 1992 – the event subsequently became men only and she could not defend her title against men – and Margaret Murdock (an American army officer and the sole woman in the competition) took the silver in the 50m rifle in 1976.[2]

Olympic shooting has come a long way. In the Paris Games of 1900, live pigeons were used as targets! The aim was to shoot as many as possible and contestants were eliminated after they missed two. More than 300 pigeons were shot. In subsequent Games, the live pigeons were replaced by the clay variety we still see today.

22

Physics for Ground Staff

Sports grounds need careful attention. Indeed, this requirement has been the death of many sports at school level. The costs of maintaining a safe grass cricket wicket, grass or cinder running track, or grass hockey pitch are extremely high. The possibility of AstroTurf, or other artificial surfaces, has alleviated the ongoing maintenance costs but requires a big initial outlay. Grass pitches are very dependent on weather, particularly sun and rain. Too much sun and the ground will become parched and hard; too much water and it will become muddy and liable to damage. How often are you going to need to water a pitch and how quickly will the water you spray, or the rain that falls, just evaporate away again?

On the average, the sun beams down a flux of sunlight of about $R = 1,366$ watts/m^2 on the earth's surface. How much gets absorbed and used to heat the surface and evaporate water depends on the type of surface it shines on. Fresh snow will bounce back 90% of the incoming light, but only about 25% of the sun's energy gets reflected back into space by the grass on a sports field, so the balance, $0.75R$, will be absorbed and will evaporate moisture from the surface layer of grass. Suppose sunlight falls on an area A and the latent heat of vaporisation of water is L. This is just the energy required to transform water into vapour without increasing its temperature. If the rate of evaporation of water from the surface area A proceeds at a rate d metres per second, then we must have:

$$0.75 \times R \times A = L \times \text{mass of water evaporated per second}$$

If the density of water is ρ, then the mass loss per second is just ρAd and we see that the area we chose, A, just cancels out. This makes sense. If the sunlight has the same intensity everywhere then the evaporation rate will be the same everywhere too and the pitch size doesn't matter (in partially covered or high-sided stadiums this might be different). This tells us that evaporation causes the level of the surface water to fall at a rate $d = 3R/4\rho L$.

The meteorologists tell us that $R = 1{,}366$ watts/m^2, the density of water is 1,000 kg/m^3, and the latent heat of vaporisation of water is measured (you do this in school science classes) to be $L = 2.5 \times 10^6$ Joules/kg and so this gives the rate of evaporation per second. It's very small so multiplied by the number of seconds in the ten hours of the day that the summer sun is shining ($60 \times 60 \times 10$s, say) we get the daily evaporation rate. You'll find that the water level evaporates at d = 1.5cm per day.

We have made lots of simplifications, as usual, in arriving at our estimates. Variations in grass quality, soil drainage, admixtures of artificial and real grass (as in most professional football pitches), wind and shade can make significant differences. But our estimate may help you decide how often to water your garden – as well as your football stadium. Four o'clock in the morning is the best time for it!

23

What Goes Up Must Come Down

If you have ever ridden on a roller coaster you will know that the greatest force you feel is at the lowest point of the ride. This is where the downward force of your weight is augmented by the downward centrifugal force you feel from your nearly circular motion around a roller coaster loop.[1] Something similar happens in the spectacular gymnastic routine on the men's high bar. When a gymnast performs a giant swing on the high bar, he has to rotate his body rigidly in a full circle while holding the bar with arms held fully extended. Only the men perform on the high bar in competition; women use the uneven parallel (sometimes called 'asymmetric') bars to perform a different range of rapid linked movements. However, women can perform a giant swing on one of the asymmetrical bars as part of their routine. In fact, Liu Xuan, who was the women's all-round Olympic gymnastics champion in 2000, became the first female gymnast to perform a *one-armed* giant swing on the asymmetric bars and this exercise is now named after her.

The giant swing requires great strength and skill to perform. The high bar is a 2.4cm steel bar raised 2.5m above the ground; leather grips are used on the hands to maintain a tight grasp of the bar. But how large are the forces that the swinging gymnast will experience when he performs a 'giant'? Like on the roller coaster, they are largest when his body is vertically below the bar.

In this configuration he feels the downward force of his own weight plus the centrifugal reaction force due to his circular motion. If the gymnast has mass M and moment of inertia I and the distance from the centre of gravity of his body to the bar is h, then we can relate his energy of rotation when his body is vertically above the bar at the top of the swing, where his angular velocity around the bar is w, to his energy of rotation when vertically below the bar, where his angular velocity has increased to W. The increase occurs because his centre of mass falls from a height h above the bar to a height h below the bar, and the consequent loss of potential energy, 2Mgh, increases the rotational energy, and hence his angular velocity, W. The energy conservation between the top and bottom of the swing is described by:

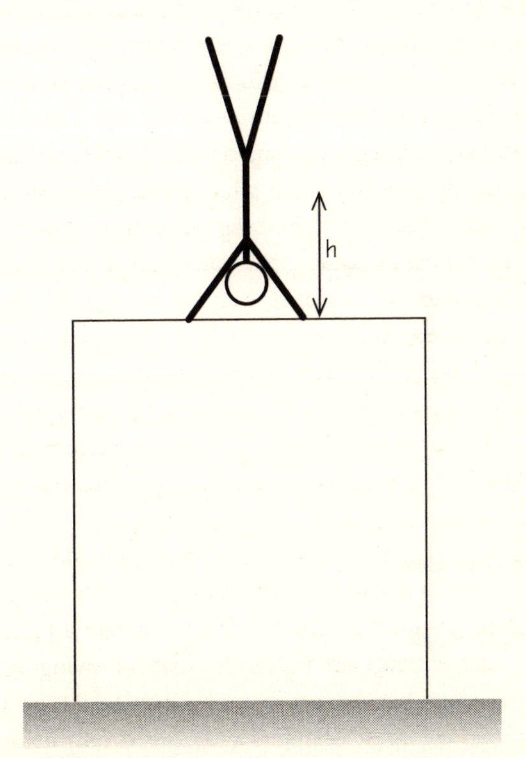

Rotational energy at the bottom = rotational energy at the top + potential energy lost:

$$\tfrac{1}{2} IW^2 = \tfrac{1}{2} Iw^2 + 2Mgh$$

At the bottom of the swing, the force on the gymnast is the sum of his weight Mg and the centrifugal force created by the rotation of his centre of gravity in a circle of radius h at an angular velocity of W, which is MhW^2; so adding these together we have:

$$\text{Maximum force felt by the gymnast} = Mg + MhW^2 = Mg + 4M^2gh^2/I$$

If we write the moment of inertia of the gymnast as $I = Mk^2$ where k is called the radius of gyration, then the total force felt becomes:

$$Mg(1 + 4h^2/k^2 + hw^2/g)$$

In practice, the last term on the right is much smaller than the second term – and if the gymnast started from a stationary handstand position, w would actually be zero. We could try to build an accurate model of the mass distribution of a gymnast's body, which is roughly a cylindrical torso with two thinner tubular legs and even thinner tubular arms attached. But, estimating roughly, the radius of gyration will be approximately the size of h, probably slightly bigger than h because the body's mass is distributed farther from the centre in the vertical direction than the lateral one. With a typical angular velocity at the top of 2–3/s, $g = 9.8m/s^2$, and a typical body size with h = 1.3m, we have approximately:

$$\text{Total force felt by the gymnast} \approx Mg(1 + 4 + 1.2) \approx 6Mg.$$

Allowing for all our simplifications and particular choices for k and h (notice that making h/k 10% smaller brings the total force nearer to 5Mg) we should simply conclude that the total force will be somewhere between 5–6Mg: that is, an acceleration between 5–6g! This stress is very severe, although the head of the gymnast is spinning in a circle with a smaller radius than h = 1.3m and so feels a smaller force.[2] Notice that even if you start from a motionless vertical handstand you only reduce it by about 1g. Interestingly, the force that a rider would feel at the bottom of a circular roller coaster ride is also about 6Mg, and to avoid such dangerous stresses roller coaster tracks are not circular. Our simple calculation reveals the extraordinary strength needed to perform even the most basic gymnastic exercise.[3] Don't try this at home!

24

Left-handers versus Right-handers

Surveys claim that about 90% of people are right-handed, about 10% are left-handed while a small number of people are mixed-handed (doing some things with one hand and other things with the other hand), and even fewer are ambidextrous: able to do all, or most things, equally well with both hands. Although there are many theories about why this imbalance persists, there is no widely accepted single explanation. People have wondered whether a small fraction of left-handers used to be at a significant advantage when swords and other hand-held weapons were employed in military activities. If you have encountered a spiral staircase in an old English castle you will notice that the spiral turns in a right-handed sense going upwards to favour right-handed defenders over right-handed attackers. The defenders have room to swing their swords inwards against the poor right-handed attackers whose sword swings keep being obstructed by the centre wall of the staircase. But if you had a special squad of left-handed storm troopers for attacking such castles then they would be much more successful than their right-handed compatriots. Alternatively, others have suggested that there might be advantages when left-handers perform tasks requiring manual dexterity and control because the left side of the body is controlled by the right hemisphere of the brain. For an interesting survey of this fascinating subject see the book *Right-Hand, Left-Hand* by Chris McManus.[1]

Let's forget about the mixed-handed and ambidextrous minorities and imagine that 90% of sports competitors are right-handed and 10% are left-handed. What will happen when right- and left-handed players encounter one another in sports like boxing, baseball, cricket, fencing or judo? The right-handers will encounter right-handed opponents in 90% of their competitions but will have the relatively unfamiliar experience of competing against left-handers in only 10% of their competitions. The left-handers on the other hand will encounter right-handers in 90% of their matches and will be better equipped from experience to outwit them than the right-handers will be to outwit them. The left-handers will have the unfamiliar experience of playing against other left-handers in only 10% of their matches but neither opponent will be at a disadvantage compared to the other. So, overall right versus right and left versus left are evenly matched but left versus right is significantly tilted in favour of the left-handers because of their greater experience of such mixed-handed matches.

25

Ultimate Pole-vaulting

Pole-vaulting had very practical beginnings. In the lowland countries of Europe and the fens of East Anglia in England, moving across farmland was continually inhibited by the network of marches, ditches and small canals that irrigated the fields and kept them drained. So every cottage kept a stock of poles for people simply to vault across the ditches. Eventually, rural sports events emerged that celebrated this skill – although the aim was to vault for distance not for height.

When pole-vaulting entered the roster of Olympic sports in 1896 (and in 2000 for women) the vaulters used poles made from ash, bamboo or aluminium, and landed on grass, sawdust or sand. The style was correspondingly simple with an onus on landing feet first and not breaking your neck! Eventually, new technologies transformed the event by introducing fibreglass and carbon-fibre poles and air-bed landing areas that freed the vaulters to exploit their formidable gymnastic abilities. Poles now come in a variety of lengths and weights, depending upon the size and strength of the vaulter, and top-class vaulters will have a number on hand to choose from during a competition. As vaulters increase in strength and body-weight they need a stiffer pole and it is important not to have one that is too weak for your weight. If your pole snaps in mid vault then you risk very serious injury, perhaps falling to the ground outside the air bed or being speared by a sharp jagged broken pole.

The men's world record is one of the best in any event. It stands at 6.14m for an outdoor vault and 6.15m for an indoor vault; both were set by Sergei Bubka in 1994. The women's record is 5.06m, set by Yelena Isinbayeva in 2009. Both performances are far ahead of the next best athletes.

Pole-vaulters sprint down the runway for about twenty strides carrying the pole with two handholds on one side of its centre of gravity. This can require the application of a force that is about five times the weight of the pole in order to keep it horizontal. Then the pole is planted at speed in the small box below the uprights that support the bar. The vaulter bends the pole, storing some of his energy of motion as elastic energy, and rotates his body about the point where the pole is planted, dropping one of his legs and rotating himself upwards. Eventually his hips will be level with his head with his body bent in an L-shape and his legs vertically upwards. The pole will be uncoiling its stored elastic energy and as it straightens out the vaulter pulls his body upwards with both arms so that it is parallel with the pole and then launches himself upwards off the end of the pole. He then twists his body over the bar with his chest facing the bar (the opposite way up to the Fosbury-Flopping high jumpers who clear the bar on their backs) so that his overall centre of gravity passes under the bar whilst his body curls over it. Landing in the air bed is best done on your back. If you land feet first you will probably twist your ankle.

A tall (1.83m) vaulter will have a centre of gravity about 1.2m above the ground. If he can sprint in to plant the pole in the box at a speed corresponding to running 100m in 10.5s (that is, $v = 9.5\text{m/s}$), then if he manoeuvres the pole most efficiently he can raise his centre of gravity by at most $v^2/2g = 4.6\text{m}$, where $g = 9.8\ \text{m/s}^2$ is the acceleration due to gravity. This shows that he can get his centre of gravity $1.2 + 4.6 = 5.8\text{m}$ above the ground. Fibreglass poles are able to add 50–90cm to a vaulter's maximum clearance height because their extra flexibility reduces the amount of the vaulter's energy that is lost bending the pole. The bend itself allows them to take off at a shallower angle, again losing less of their stored energy. The pole flexibility is therefore not being used simply as a catapult to launch them in the air.

The post-Bubka super-vaulter of the future might be able to realise that optimal use of his take-off energy and then use their strength to pull themselves up and launch from a vertical handstand position with arms off the end of his pole. This would raise his centre of gravity by a further distance of about 1.8m (two arms' lengths) and he would be looking at a possible clearance of 7.6m. A key factor in the whole sequence is the take-off speed achieved carrying the pole because the square of that speed determines how high his centre of gravity will initially go. Strength plays the major role in determining that speed and also governs the ability to bend the pole and then raise the body up high above it. Of course, we are assuming that no major technological advance creates a new type of lighter, more dynamic pole that does for fibreglass what fibreglass once did for bamboo.

26

The Return of the Karate Kid

There are two martial arts in the Olympic Games, the Japanese sport of judo* (introduced in the Tokyo Games of 1964 for men and in Barcelona 1992 for women, and in the Paralympics in 1988) and the South Korean sport of taekwondo† (introduced in the Sydney Games of 2000 for men and women after being a demonstration sport in 1992). It is surprising that karate, the third major martial art which is practised worldwide, isn't an Olympic sport. Its inclusion was discussed as far back as 2001, but the formal proposals to the Programme Commission of the IOC in 2005 and 2009 both failed.[1] It has been reported that karate lost out because of the different rules and styles that exist around the world – some fight with light contact, others with full contact, some with or without protective padding, others with boxing gloves – and it was not clear how to coordinate all this fairly into a single agreed Olympic competition. There are at least a dozen officially recognised styles and associations. Olympic karate would have to choose a single style and this would mean that lots of competitors would have to change what they had learned and would be at a disadvantage against those who didn't need to. Bringing them all together was compared with the politics of merging different branches of a major religious faith into a single church.

*'Judo' means 'the way of gentleness'.

†'Tae' means to break or strike with the foot, 'kwon' to do the same with the fist, and 'do' is just the art of so doing.

The most dramatic exhibition of karate is traditionally the demonstration of a single chop breaking a brick or a wooden plank. Let's see how easy this is for a black-belt exponent. The key to an effective karate strike is speed and acceleration in the final moments before a blow strikes its target. A top black belt can deliver a blow at about 7m/s. The average mass of a man's arm is about 3.4kg so the momentum of the striking arm is about $7 \times 3.4 = 24$kg m/s and the contact time with the target is less than five milliseconds, so the force exerted can be as large as $24 \div 0.005 = 4,800$N. For comparison, a 70kg person exerts a force of $70 \times 9.8 = 686$N on the ground. If this impact force should strike your head, which has a mass of about 5kg, then the resulting acceleration would be $4,800/5 = 960$ m/s^2, or about 96g.[2]

How does this compare with what is needed to split the plank or the brick? You will notice that karate demonstrators who attack planks will generally use a stack of thin planks to make up the total thickness rather than a single thicker plank. This is an easier challenge because you just need to break one thinner plank after the other rather than split the single stronger plank. The continued downward momentum of the split planks at the top of the stack help break the ones lower down: breaking two planks requires less than twice the force needed to break one. Also, in demonstrations notice that the planks will be orientated so that the hand-strike is parallel to the grain of the wood where it is easiest to split it and it is important to hit the stack of planks right in the centre.

The expert will be able to concentrate so keenly that his blow will be coming in at maximum speed when impact occurs. If you get nervous about hurting your hand and slow it down then the force exerted will drop dramatically – and you will hurt your hand even more because the wood won't break.[3] About 3,100N of force will be required to break a 20cm × 30cm pine plank that is 1cm thick, and about 3,200N to split a brick of the same area that is 4 cm thick. So you can see that the 4,800N of force that our black belt can apply will easily break a brick or a stack of half a dozen planks.

27

Leverage

We are all familiar with levers. A large force can be created by exerting a force a long way away from its point of application. The key quantity is the product of the force applied and the distance from the point where equilibrium is possible, called the fulcrum. In the pictures below you can see three different types of lever and the balance of forces that is entailed. The so-called 'Class 1' lever has a load applied on one side of the fulcrum and the downward force needed to balance it is applied on the other side. If it is applied at the same distance from the fulcrum as the load, then it will have to be equal to the load's weight in order to be balanced. But if it is applied at a greater distance from the fulcrum than the load, it can be smaller. Balance is obtained when the product of the load force times the distance from the fulcrum is the same on both sides.[1] This type of lever is familiar. If you sit on a see-saw with someone of the same weight, equidistant from the centre, then this type of balance results.

The 'Class 2' lever is different. The effort is applied farther from the fulcrum than the load, on the same side of the fulcrum and applied in the opposite direction. If someone sits on one side of a see-saw and you try to lift them up by raising their seat from behind then you are applying this class of leverage. Finally, the 'Class 3' lever reverses the positions of the load and the effort and you are trying to lift a load that is farther away from the fulcrum – this is rather more strenuous.

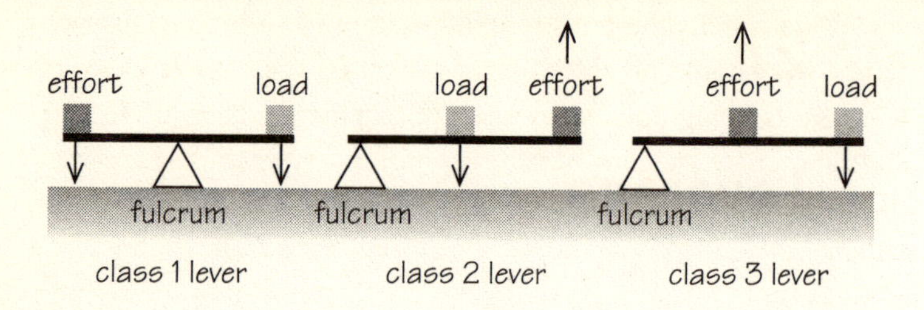

Applications of these three varieties of lever crop up all over the Olympic sporting schedule and in the training programmes that athletes use. If you try to do some press-ups then you are operating as a Class 2 lever. Your toes are the fulcrum; your body weight is the load and you are using your arm muscles to apply force upwards in order to overcome it. If you sit down and hold a barbell at arm's length before curling it up towards you then you are creating a Class 3 lever. Do a sit-up with your feet held down on the ground by someone else and you are working as a Class 3 lever also. Rowers are using their oars as Class 1 levers with the rowlock as the fulcrum as they pull them through the water.

The more interesting (and painful) applications to watch for are in the wrestling events. Olympic wrestling should not be confused with TV wrestling – the Dark Destroyer, Giant Haystacks, and all those familiar entertainers – and is a much slower and more tactical battle as the protagonists attempt to make one of the five types of move – a takedown, a breakdown, a pin, a reversal and counter – or an escape. In the Olympic Games you see 'free-style' and 'Greco-Roman' style, which have different rules. In Greco-Roman you can't make holds below the hips and use of the legs is not allowed. In freestyle there are still restricted holds, but use of the legs and holds below the hips are allowed.

In all these wrestling events it is the holds applying Class 2 leverage that are regarded as the most effective. Those that act like Class 3 levers are the least efficient and those of Class 1 are in between.

As you watch different strength events across the sporting spectrum, you should be able to spot which of these three types of leverage is being applied by the participants as they attempt to shift inanimate weights, or other competitors, by judicious choice of fulcrum and their own considerable strength and powers of balance.

28

Reach for the Sky

In rugby union there is a way to reintroduce the oval ball back into play that is more structured and dramatic than the simple throw-ins used in soccer. The lineout sees two rows of players await the delivery of the ball by a member of the team that didn't last knock it out of play. The waiting players don't only jump, some of them are launched into the air by their teammates so as to catch or palm the ball back to their waiting colleagues. Clearly, when it comes to lineout jumping, the taller you are the better: teams with no tall players are going to lose the ball at every lineout. But being tall and able to jump high from a standing position (no run-ups here) is not enough to win the ball because you are in competition with equally huge players who are being launched into the sky by pairs of their teammates after the ball has been thrown. Have you ever wondered how high they go? Two big strong players will grasp one leg each of the jumping lineman going for the incoming throw and lever him upwards. They will start with their arms roughly horizontal but by the end of their launch manoeuvre they will be almost vertical.

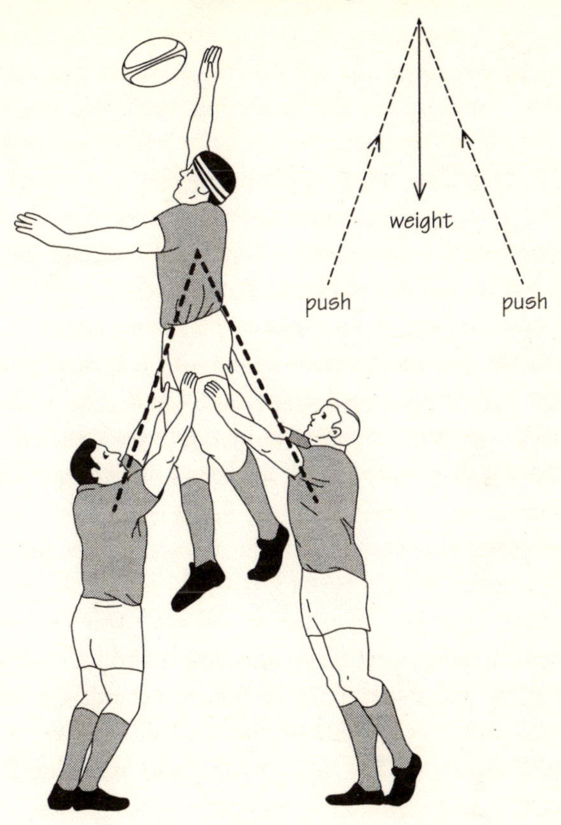

Most fit athletes can exert a force that is at least as large as their body weight (top weightlifters can lift much more), so the two lineout props will be applying an upward force of at least two of their considerable body weights to the jumper. Assuming all three of them have the same weight, the downward force of gravity on the jumper is one body weight, so he ends up feeling a net upward force of at least one body weight instead of only a net downward force of one body weight if he had no assistance. It is not uncommon to have lineout men who are 2m tall and they can stretch their arms out another three quarters of a metre above their heads. If they grasp the jumper round his thighs then he is going to be able to catch a ball 4.5–5m above the ground. That is impressive: you can win the men's pole vault gold medal at the Olympics with a clearance of 6m and the world high-jump record is 'only' 2.45m. It's a long way down and there's no air bed landing area! The force exerted means that the jumper goes up to the maximum height in about three quarters of a second. The player who throws the ball in from the touchline has to do it very precisely so that its arcing trajectory reaches a maximum height of 4.5–5m at exactly the maximum aerial position of the jumper in the line. His team will have a signal to tell him which man in the line is going to be his target jumper but he hasn't got much more than three quarters of a second to get the ball to that key spot in the air.

29

The Marathon

The men's marathon is traditionally held on the last day of the Olympic athletics programme over the curious distance of 26 miles 385 yards (that is 26.22 miles or 42.195km, if you prefer). It was part of the Olympic Games right from the start of their modern reinstitution in 1896. Those new Games were to be held in Athens, as tradition required, and the Greeks naturally wanted to create a dramatic event which resonated with Greek history. The resonance they chose to amplify was not one created by a sporting event at the ancient games but one enshrined in legend. Without delving into the arguments that historians have enjoyed about the truth of all this, the cue was taken from a story told by ancient historians about a messenger, Pheidippides (although one later writer names him as Philippides), who was sent from the battle at Marathon to Athens with news of the Persians' defeat in August or September 490 BC. The terrain was hilly and rugged and the various routes he might have chosen varied in length from about 23 to 26 miles, although back in 1896 it was assumed that he took the longer distance. When he arrived in Athens he announced the good news 'We have won' and then promptly collapsed and died.

The winner of the 'Marathon' race in the 1896 Athens Games (in 2hr 58m 50s) was an unknown Greek water carrier named Spiridon Louis, although curiously he had only come fifth (in 3hr 18m) in the trial race over the same course that was employed to select the Greek team. A photo of the leaders is shown here.

If you are running a marathon for fun you might be impressed with that winning time. Alas, its not quite what it appears. The distance used today to define the marathon was not standardised by the IAAF until May 1921 and the actual distances run in those early marathons are shown in the table here:

Year	Distance (km)	Distance (miles)
1896	40	24.85
1900	40.26	25.02
1904	40	24.85
1906	41.86	26.01
1908	42.195	26.22
1912	40.2	24.98
1920	42.75	26.56
1924 onward	42.195	26.22

No one was too bothered about the actual distance run in the early Olympics; all that mattered was that the competitors ran the same distance in each single race. The 1904 marathon at the St Louis Olympics was like something out of the Keystone Cops. The race was run on roads amongst traffic, which the runners had to dodge all the time, and one runner was chased off the course by dogs. The first competitor to finish was subsequently disqualified after it was revealed that he had travelled nearly half of the race by car. As a result of these vagaries, the distances varied between 24.85 and 26.56 miles up until the 1920 Games, since when the new standard distance of 26 miles and 385 yards first run in 1908 has been used for all men's, women's and Paralympic marathons. Where did this standard distance come from?

We owe the official marathon distance to the British Royal Family. When the 1908 Games were held in London, the athletics events took place at the old White City stadium. The Royal Family asked if the marathon race could start outside the gates of Windsor Castle so that their children could witness it from the windows of the nursery; the event then also shifted the finish line on the White City track so that it was in front of King Edward VII in the Royal Box. These two concessions combined to produce the now iconic distance of 26 miles and 385 yards.

The 1908 Olympic marathon race was very dramatic. It finished in the stadium in front of 68,000 spectators and the leading runner, the Italian Dorando Pietri, entered the stadium in an extremely distressed state, collapsing on the track and becoming increasingly disorientated. Two officials near the finish took pity on him, held him by the arms and walked him over the finish line. Alas, he was immediately disqualified. The gold medal was taken by the 22-year-old American Johnny Hayes, one of many Irish-born athletes competing for the United States, who completed the course in 2hr 55m 18s.[1] But nobody remembers Hayes. Pietri's fate so impressed the Royal Family that Queen Alexandra presented him with a special gilded silver cup the following day. Over the next four years both Hayes and Pietri turned professional and met on four occasions in races over the marathon distance: Pietri won each time. Many years later, the English runner Joe Deakin ascribed Pietri's disorientation and collapse to the fact that he had drunk quite a lot of the alcohol that spectators had been offering him along the course.

30

All That Glitters Is Not Gold

For most sports competitors an Olympic gold medal is the pinnacle of achievement. Yet in the ancient Greek Olympics there were no medals. The winner of each event received an olive wreath and the second- and third-place competitors got nothing. When the Olympics were revived in 1896 the winner received a silver medal and the runner-up a bronze medal.[1] Curiously, gold seems to have been thought of as inferior to silver at the time. The 1900 Olympics in Paris did not even award any medals, just cups and other trophies – and the winner of the men's 200m freestyle swimming, the Australian Frederick Lane, was awarded a bronze statue of a horse! Then, in 1904, pure gold medals replaced silver for the winners, and silver and bronze medals were also awarded. Yet pure gold medals disappeared after 1912 when Olympic 'gold' medals were made of sterling silver and coated with 6g of gold to provide a spectacular golden colour. They were required to be at least 60mm in diameter and at least 3mm thick.

Each host nation is responsible for the manufacture of the medals presented at their Games. Throughout the whole of the period from 1928 to 1968, they had the same two designs on each face, which were created by the Italian artist Giuseppe Cassioli. After 1972, there was increasing latitude for the hosts to apply their own design to one side of the medal. At the Vancouver Winter Games in 2010 much publicity was given to the fact that the medals were eco-friendly because they had been made from metals that had been recycled from old TVs and computer circuits. In Beijing in

2008, they were 70mm across and 6mm thick, weighing 120g, and contained a ring of rare light jade as well as gold. In London in 2012, the gold medals will be the largest ever awarded at a Summer Games, 85mm across, 7mm thick, and weighing 400g.

31

Don't Blink First

Taking penalties is a nerve-wracking business at any time, but the penalty shoot-out that decides the outcome of the whole football game is likely to have long-term consequences for anyone who misses. What is the best strategy? A random placing of the ball (between the posts and under the bar of course) is a tactic that no opponent can beat with a definitive strategy. But there may be a non-random element that can be exploited. Individual goal-keepers and penalty takers will have strong and weak sides – depending, for instance, on whether they are right-handed or left-footed. They may be weaker at getting down to low shots if they are very tall, and liable to blast the ball over the crossbar if they are defenders used to simply banging the ball into row Z of the stand.

Dedicated researchers have studied the taking of penalties in football matches all over Europe.[1] Between them, they gathered data on where the taker placed the ball and in which direction the goalkeeper dived for 1,417 different penalties. Dividing the options for both the kicker and the goalkeeper into three choices for them to shoot or dive – to the right, to the left, or stay central – the percentages of successfully taken penalties ('successful', that is, from the taker's point of view because a goal was scored) is shown in this table:

	Penalty Taker		
Goalkeeper	Shoots left	Shoots centre	Shoots right
Dive left	60	90	93
Stay in centre	100	30	100
Dive right	94	85	60

You notice the preponderance of goalkeepers' dives to the right in preference to the left, which is surely a reflection of the fact that roughly 90% of the population at large are right-handed. In practice, both goalkeepers and penalty takers can use these percentages to work out an optimal strategy for taking and saving penalties. The optimal strategy is the one that minimises the worst that can happen to you and there is a simple and well-studied mathematical theory to extract what this strategy is for players in a two-person game. It was created by the famous mathematician John Nash, whose life became the subject of the award-winning film *A Beautiful Mind*. Nash eventually received the Nobel Prize for economics in recognition of this important work. The 'Nash Equilibrium' for the contest between the penalty taker and the goalkeeper is that the kicker should put 37% of his shots to the left, 29% down the middle and 34% to the right. The goalkeeper should adopt a best strategy of diving left 44% of the time, staying in the middle 13% and diving right 43% of the time.

If both adopt their optimal strategies then about 80% of penalties will be scored.[2] So in a typical shoot-out with each team taking five each you expect two to be missed if all are taken with these optimal strategies.

32

Ping-pong is Coming Home

Table tennis may be played by more people than any other sport in the world. Just about everyone in China seems to play the game. When I visited, I found weatherproof outdoor tables in the most unlikely public places and always a queue of players and lines of keen spectators. The most accomplished sportsman or woman to pass through my own university is the remarkable Chinese table-tennis player Deng Yaping, with four Olympic gold medals for singles and doubles in 1992 and 1996 and six world championships (singles and doubles in 1991, 1995 and 1997), who is regarded as one of the greatest ever players and is the most famous sporting figure in China.

Table tennis became an Olympic sport in 1988 and is one of those games that has changed its scoring system quite recently in order to make it more exciting for spectators. Originally players had to score at least 21 points with a 2-point margin and took it in turns to serve for 5 points; the winner was the best of three games. Now, top-level table-tennis games are won by scoring at least 11 points with a 2-point margin and the best of seven games.[1] Each player now only has two serves except after 10–10 when they take a single serve each. These rule changes stop the server holding this advantage for so many points and also prevent long one-sided games.

Ideally, scoring systems should try to increase the number of times that either player is on the verge of winning the match and minimise the chance that the weaker player wins as a result of pure luck. For example, if the winner was decided by playing 1 point, then there is less opportunity for the more skilful player to assert himself. However, while playing lots of points will make it harder for the weaker player to hit lucky time and time again, it can have a negative effect on spectators' enjoyment. This is why we see sequences of games and sets in many sports. It keeps the game alive for longer, stops irretrievable points gaps being created and produces lots of key points on which games, sets, and ultimately the match, hinge.

If one player has the same probability, p, of wining each point and you need to win n games to be the winner of the match, then we can calculate the chance of this player having n successes before having n losses. For simplicity, suppose that the two players are very evenly matched so p is very close to ½. Writing $p = ½ + s$, where s is a very small quantity (much less than ½), the probability of n wins before n losses is approximately:[2]

$$Q = ½ + 2s\sqrt{(n/\pi)}$$

Note that if the chance of winning each point is exactly ½, then s is zero and the match-winning probability, Q, is ½ also. We can also see that a very small advantage for one player (s very small and positive) gets steadily amplified (growing as the square root of n) as the number of games played increases. The more evenly matched the two players, so the more games you need to play in order for Q to become significantly larger than ½. This is why, for example, top-level men's tennis matches originally used five sets rather than just three.

The general rule is that for evenly matched players, the marginal advantage grows as $2s\sqrt{(n/\pi)}$ in the quest to be the first to win n games. Roughly, the probability of winning a game to n points by at least 2 clear points and then winning m games to win the whole match will be $2\sqrt{(m/\pi)} \times 2s \sqrt{(n/\pi)} = (4s/\pi)\sqrt{(m \times n)}$.[3]

The key quantity here is $m \times n$. If $m \times n$ is the same for two different scoring systems then they will be equally rewarding for skill over luck for evenly matched players. What is the case with the table-tennis scoring changes? In the old days they needed $n = 21$ points to win a game and $m = 2$ games to win the match because they played best of three. So, $m \times n = 42$. Now we play $n = 11$ points and play best of seven, so $m = 4$ and $m \times n = 44$. The numbers 42 and 44 are so close that somebody must have known what they were doing when they changed the table tennis scoring system! The new system is almost the same in terms of its reward for skill over luck but offers twice as much spectator interest and less server bias.[4] The common factor is fairly clear: $m \times n$ is a good measure of the total number of games that are going to be played in the match. Keeping this roughly the same is useful for planning purposes (and TV scheduling) while it also maintains the skill-to-luck reward factor.

33

A Walk on the Wild Side

Race walking, under the title 'pedestrianism', was included in the first athletics meeting of the English Amateur Athletics Association in 1880, and in the Olympic Games of 1904 it was part of a decathlon-style multi-event challenge. (What an interesting thought, to add the walk to the modern decathlon!) Individual walking races were introduced in 1908 and have been part of the Olympics ever since, with both 20km and 50km road events. Women's events were added in 1992 and feature the 20km race only. The world record for the men's 20km on the road is currently 1hr 16m 43s. For women it is 1hr 24m 50s. This is an average speed of 4.35m/s over 12.5 miles, and at 6m 8s per mile this is a very respectable running pace. This is clearly not walking as we know it on an everyday perambulatory basis.

How do race walkers go so fast? The rules place considerable constraints on them. They must have one foot in contact with the ground at all times, otherwise they risk disqualification for 'lifting', and the contact leg must be straight (not bent at the knee) when it is supporting the body weight in a vertical position. These rules define the difference between walking and running. When you run you will lose contact with the ground each time you drive off from the ground and your leg will generally be bent during this movement. In practice, race walkers generate their remarkable speeds by an extremely high rate of cadence. Top walkers will be pacing at the same rate as a 400m runner but keeping it up for seventy-five minutes, not just forty-five seconds. The speed they can move at will be determined by the number of strides per second times their stride length. If they walked like the ordinary pedestrian in the street then their stride length would be too short to generate the high speeds required. This is why top-class walkers display that distinctive hip-swinging movement: it effectively extends their stride length. This hip-extension technique requires great flexibility and coordination and the leading exponents walk with extraordinary smoothness. Whereas the centre of gravity of a runner will move up and down as they run through a sequence of strides, the centre of gravity of a top walker will move forwards in a straight line. No energy is wasted moving the centre of gravity up and down unnecessarily.

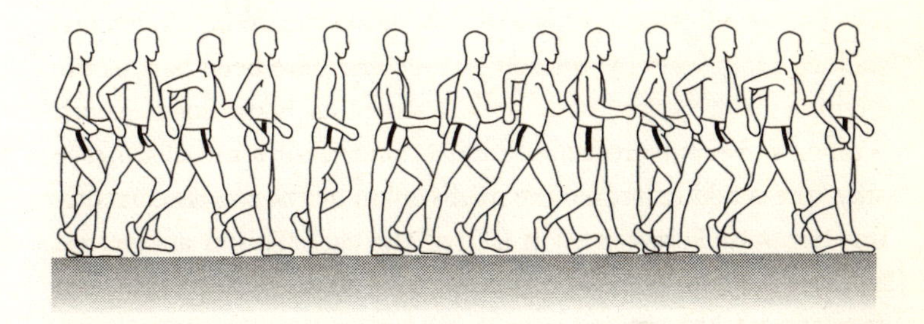

Race walking has a big problem, however. Its rules were devised in the era before television, and even now they are only enforced by human judges who determine whether or not walkers are 'lifting' at widely spaced intervals on a very long road course away from the stadium. Close scrutiny by TV cameras showing slow-motion replays has led to an outcry from journalists and viewers that all the top walkers are running. Contact is constantly being broken and the event is said to be turning into modified running. However, take a look at film of the remarkable Jefferson Perez of Ecuador, the 1996 Olympic champion.[1] Remarkably, this film is not speeded up – he really is taking 186 paces per minute – and his style is perfect: there is no up-and-down motion, and the still frames seem to suggest that he is not lifting.

It is interesting to subject the contact to a more mathematical analysis.[2] At the moment when a walker has both feet in contact with the ground his legs of length L form a triangle. His stride length is S.

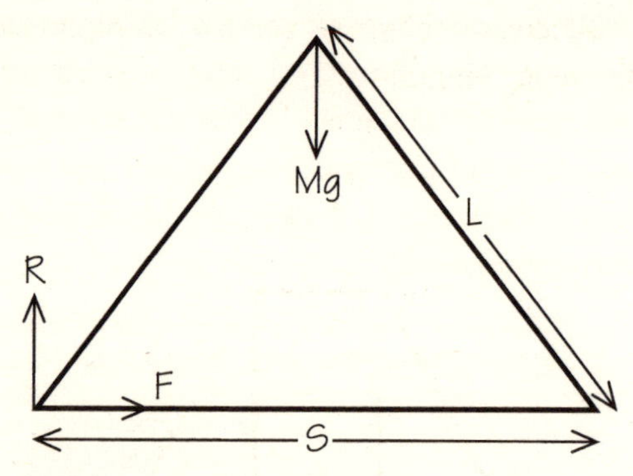

The forces acting in the vertical direction are the walker's weight and the upward reaction force, R, of the back foot on the ground. If that back foot breaks contact with the ground before the front foot touches the ground then R will become zero and the walker will be 'lifting'. The maximum speed, V, that can be achieved without lifting is given by the formula:

$$V^2 = \tfrac{1}{2}\, gL[3\sqrt{(4 - S^2/L^2)} - 4]$$

where $g = 9.8m/s^2$ is the acceleration due to gravity, L is the length of the walker's leg and S is the length of the walker's stride. If we assume a typical leg length of 1m and a stride length of 1.3m then the maximum walking speed that can be achieved without breaking contact with the ground is V = 1.7m/s. This is woefully short of the speeds in excess of 4m/s that top walkers sustain over 20km.

There are two things that might make them go faster without

lifting. The first is to use a shorter stride length (smaller value of S) with a faster cadence. For example, suppose we use S = 1m for the stride length, then we can increase the maximum legal walking speed to 2.42m/s. But this is not enough, so let's look at the second key factor – the leg length of the walker. By adopting the characteristic hip sway a top walker can increase the effective length of his legs. This increases the stride length by about the same amount as the effective leg length.

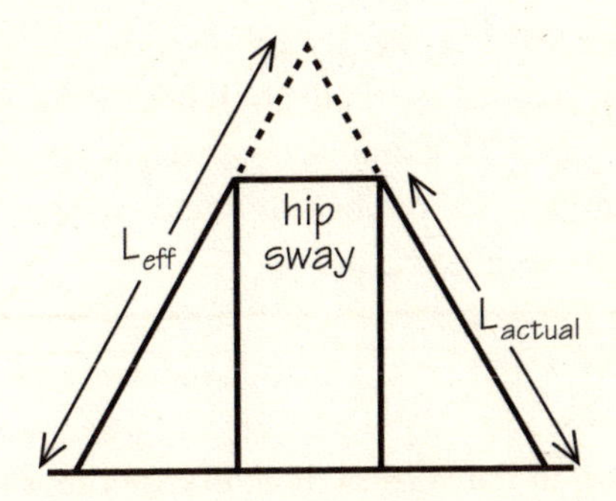

With the new effective leg length equal to the stride length it is only possible to achieve world record speeds of 4.35m/s without lifting if the hip motion leads to an effective leg length of about 2.3m, which is absurd. This does suggest that contact is continually being broken at these high walking speeds.

34

Racing Certainties

I once saw an episode of the TV crime show *Midsomer Murders* that involved a plan to defraud bookmakers by nobbling the favourite for a race. The drama centred on other events, like murder, and the basis for the betting fraud was never explained. What might have been going on?

Suppose that you have a race where there are published odds on the competitors of a_1 to 1, a_2 to 1, a_3 to 1, and so on, for any number, N, of runners in the race. If the odds are 5 to 4 then we express that as an a_i of 5/4 to 1. If we lay bets on all of the N runners in proportion to the odds, so that we bet a fraction $1/(a_i + 1)$ of the total stake money on the runner with odds of a_i to 1, then we will always show a profit so long as the sum of the odds, which we call Q, satisfies the inequality:

$$Q = 1/(a_1 + 1) + 1/(a_2 + 1) + 1/(a_3 + 1) + \ldots + 1/(a_N + 1) < 1$$

And if Q is indeed less than 1, then our winnings will be at least equal to:

$$\text{Winnings} = (1/Q - 1) \times \text{our total stake}$$

Let's look at some examples. Suppose there are four runners and the odds for each are 6 to 1, 7 to 2, 2 to 1 and 8 to 1. Then we have $a_1 = 6$, $a_2 = 7/2$, $a_3 = 2$ and $a_4 = 8$ and:

$$Q = 1/7 + 2/9 + 1/3 + 1/9 = 51/63 < 1$$

So, by spread-betting our stake money with 1/7 on runner one, 2/9 on runner two, 1/3 on runner three, and 1/9 on runner four, we will win at least 51/63 of the money we staked (and of course we get our stake money back as well).

However, suppose that in the next race the odds on the four runners are 3 to 1, 7 to 1, 3 to 2 and 1 to 1 (i.e. 'evens'). Now we see that we have:

$$Q = 1/4 + 1/8 + 2/5 + 1/2 = 51/40 > 1$$

and there is no way that we can guarantee a positive return. Generally, we can see that if there is a large field of runners (so the number N is large) there is a better chance of Q being greater than 1. But large N doesn't necessarily guarantee that we have $Q > 1$.[1]

But let's return to the TV programme. How is the situation changed if we know ahead of the race that the favourite in our $Q > 1$ example will not be a contender because he has been doped?

If we use this inside doping information we will discount the favourite (with odds of 2 to 1) and place none of our stake money on him. So, we are really betting on a three-horse race where:

$$Q = 1/4 + 1/8 + 2/5 = 31/40 < 1$$

and by betting 1/4 of our stake money on runner one, 1/8 on runner two, and 2/5 on runner three we are guaranteed a minimum return of $(40/31) - 1 = 9/31$ of our total stake in addition to our original stake money! So we are quids in.

Incidentally, even when $Q > 1$ it is reputed that the system is useful – for money laundering. If you bet on all the horses in the manner presented then you get your cash laundered by the bookmakers but there is a 'charge' that is $(1/Q - 1)$ times the money staked.

35

What is the Chance of Being Disqualified?

There were a number of high-profile disqualifications at the 2011 World Athletics championships. When Usain Bolt was eliminated from the final of the 100m there was a huge groan from the crowd, the meeting organisers and the commercial sponsors. The new(ish) false-start rule gives no one a second chance: if you false start once then you are disqualified (DQd). The old false-start rule was more athlete-friendly. If anyone false started they were given a second chance but the next false starter was DQd regardless of what they had done already; in short, if you committed the second false start you were out. Some people didn't like this because they claimed that it encouraged gamesmanship (and even gameswomanship): weaker starters could false start deliberately so as to make their faster-starting rivals more cautious next time. But the real reason for the new rule was the TV coverage. False starts throw the event behind schedule and irritate impatient programme producers. 'One strike and you are out' minimises the potential disruption to the programme schedule.

Suppose that there are n runners in the 100m final (n = 8 usually) and they have probabilities of false starting that are equal to p_1, p_2 . . . p_n respectively. Under the current rules, the probability of any athlete (r, say) being DQd is simply p_r, for r = 1, 2, 3 . . . n. Also, it is possible for more than one athlete to be DQd at once.

Under the old rules there were many more ways for athlete r

to get DQd. They could false start twice (with probability $p_r \times p_r = p_r^2$) or they could be the second person to false start after any of the other runners had false started. If runner 1 false started before them then this probability would be $p_1 \times p_r$ because we assume them to act independently. So to calculate the total probability of being the second athlete to false start, and so be DQd, we just add up these probabilities for following a false start by every one of the *other* runners. The total probability that r will be DQd under the old rules is therefore $p_r(p_1 + p_2 + p_{r-1} + p_r + p_{r+1} + \ldots + p_n)$.

The second factor (in the brackets) is the sum of the probabilities that each runner will false start and so measures the likelihood of a false start – it is a measure of the 'nervousness' of the field. If its value exceeds 1 then the chance that r will be DQd was greater under these old rules than under the new ones. But if the 'nervousness' total is less than 1 then the chance of being DQd is lower overall under the new rules and the TV schedule is safer. Notice, however, that the relative probability that two given athletes, r and s, will be DQd is the same under both rules, and equal to p_r/p_s, so the rules don't favour people who are more or less likely to false start. As a concrete illustration suppose that in a race with eight competitors they all have an equal chance of 1/32 of false starting. Under the new rules each competitor has a 1/32 chance of being DQd but under the old rules that chance was only $1/32 \times (8 \times 1/32) = 1/128$.

36

Rowing Has Its Moments

If you look at the pattern of rowers in a racing four or eight then you expect to find them positioned in a symmetrical fashion, alternately right – left, right – left as you go from one end of the boat to the other. This pattern is called the 'rig' of the boat and the one we have just described is called the 'standard rig', shown here for a four and an eight:

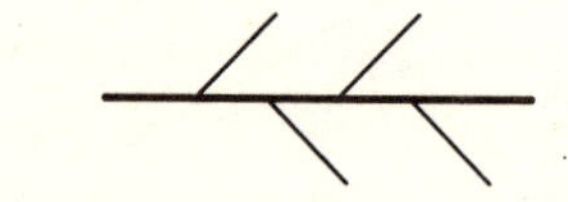

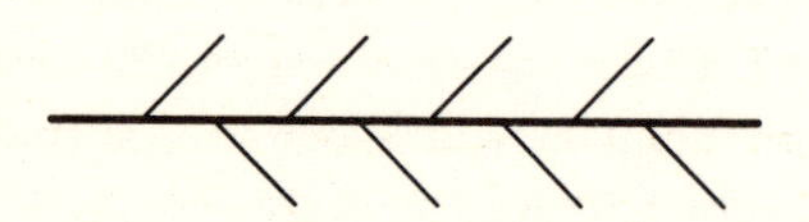

However, the regularity of the rower's positions hides a significant asymmetry that affects the way the boat will move through the water, which I examined a few years ago.[1] If we look at the rower pulling the oar towards him, and then at the rowlock holding the oar, the total force acting on the boat can be split into two distinct parts: the component parallel to the boat in the direction of the boat's forward motion and a component at right angles to it. The interesting force is the second one. During the first half of the stroke it is directed towards the boat (force F) but during the second 'recovery' phase the force is in the opposite direction (force F'), at right angles to the boat but away from it. As a result the boat is subject to a sideways force that alternates, first towards the boat, then away from it.

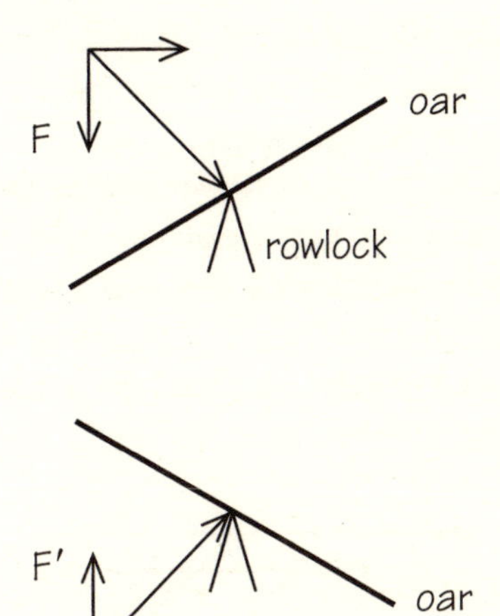

If we go to the back of the boat we can take moments of these sideways forces. Each is equal to the magnitude of the force times the distance to the point of application of the force. For simplicity, let's assume the rowers have identical strengths, equal to F, and the first rower (the 'stroke') is at a distance s from the end of the boat and the other rowers are equally spaced, a distance r apart. During the first half-stroke the sum of the moments of the transverse forces acting on the four rowers in a four is:

$$M = sF - (r + s)F + (r + 2s)F - (r + 3s)F = -2Fr$$

Notice that the answer is not zero! Also, conveniently, the distance to the first rower, s, cancels out because there are the same number of rowers on each side of the boat. In the second phase of the stroke everything is the same except that the direction of the force F reverses. This simply means changing F to –F in our formula. As the boat is propelled forwards it is subjected to an alternating sideways force that varies between +2Fr and –2Fr: it *wiggles*.

If there is no cox the rowers have to discern this wiggle and expend energy cancelling it out by sideways countermovements. If there is a cox then they will use the rudder to counter the wiggle. Both actions use up energy.

We can avoid the wiggle by repositioning the rowers. In the case of the four, a rig that goes right – left – left – right, as shown below, results in no net sideways moment acting on the boat:

$$M = sF - (s + r)F - (s + 2r)F + (s + 3r)F = 0$$

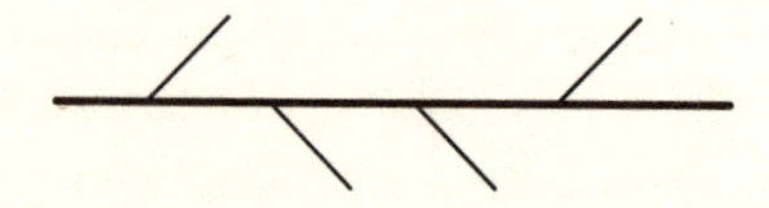

This seating plan is known as the 'Italian' rig because it was discovered by the Moto Guzzi Club team on Lake Como in 1956. The crew of the club's four was being watched by Giulio Cesare Carcano, one of the company's leading motorcycle engineers, who suggested that the failure of the boat to run straight might be alleviated by putting the middle two oarsmen both on the starboard side.[2] The result was so successful that the Moto Guzzi crew went on to represent Italy and take the gold medal that year at the Melbourne Olympics.

The situation for an eight is more complicated. However, the standard rig produces a non-zero sideways moment that alternates between –4Fr and +4Fr during each stroke. The cox is going to have to counter that wiggle. But I found that there are four possible ways that an eight, crewed by identical rowers, can have a zero sideways moment.[3] Here are the four no-wiggle rigs:

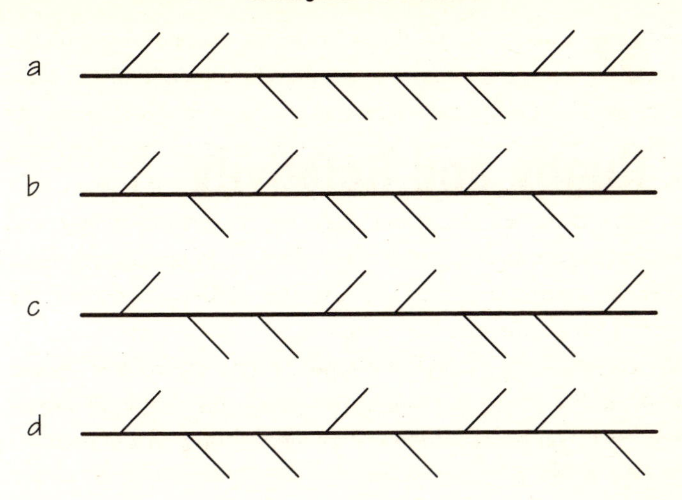

Rig (c) is just two of the Italian rigs for a four set one behind the other. Rig (b) is the so-called 'German', 'bucket' or 'Ratzeburg rig', first used by crews training at that famous German rowing club in the late 1950s under Karl Adam, who was motivated by Carcano's configuration for the four. The other two rigs are new.

When I published these results they attracted a lot of attention around the world triggering articles in *World Rowing Magazine*; the results were then confirmed in greater detail in an article in the *Rowing Biomechanics Newsletter.*[4] Subsequently, the *New Scientist* commissioned some trials on the River Thames with the Imperial College eight to see how they liked the new rig (a) compared to the standard one. The winning Canadian men's eight at the Beijing Olympics used the German rig (b) in their gold-medal race[5] and Oxford did the same to win the 2011 University Boat Race on the Thames – the first time a non-standard rig had been used in the race for forty years. Perhaps someone will try my (a) or (d) in London.

37

Rugby and Relativity

In 2003, I spent two weeks visiting the University of New South Wales in Sydney during the Rugby World Cup. Watching several of these games on television I noticed an interesting problem of relativity: what is a forward pass relative to? The written rules are clear: a forward pass occurs when the ball is thrown towards the opposing goal line. But when the players are moving the situation becomes more subtle for an observer to judge due to relativity of motion.

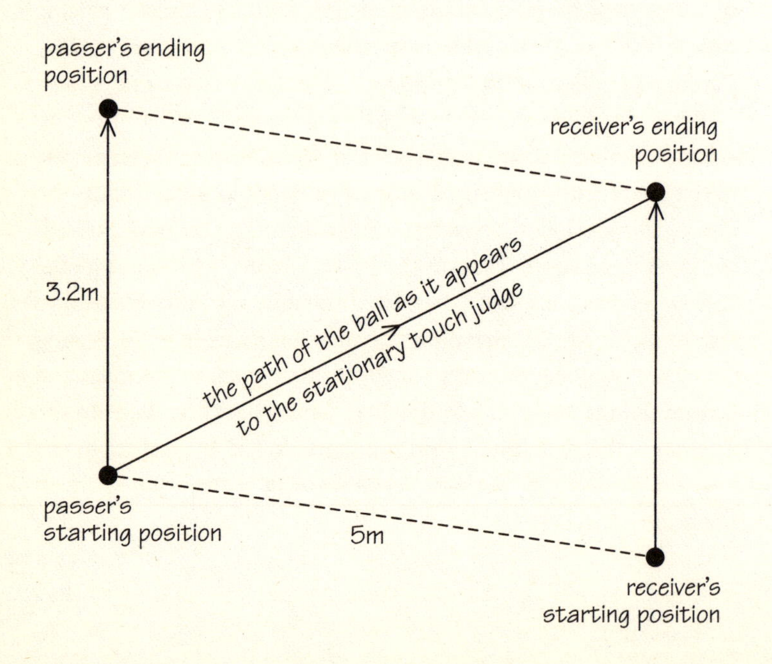

Imagine that two attacking players are running in parallel straight lines 5m apart at a speed of 8m/s towards their opponents' line. One player, the 'receiver', is a metre behind the other, the 'passer', who has the ball. The passer throws the ball at 10m/s towards the receiver. The speed of the ball relative to the ground is actually $\sqrt{(10^2 + 8^2)} = 12.8$m/s and it takes a time of 0.4s to travel the 5m between the players. During this interval the receiver has run a further distance of $8 \times 0.4 = 3.2$m. When the pass was thrown he was 1m behind the passer but when he catches the ball he is 2.2m in front of him from the point of view of a touch judge standing level with the original pass. He believes that there has been a forward pass and waves his flag. But the referee is running alongside the play, doesn't see the ball go forwards, and so waves play on!

38

Run Rates

During big cricket matches the BBC website runs a text commentary service alongside the usual scorecard and other pieces of match statistics. The most interesting of these is the graph that shows the scoring rate. In one-day cricket each team is allowed to bat for no more than 50 overs. There are a variety of views as to whether it is best to choose to bat first or second if you win the toss before the match starts. But, whoever bats, the scoring rate is important. It is no good having an immovable batsman in your team who scores just one run an over and blocks progress – you'd probably want your other batsman to try to run him out so as to let in a faster scorer! Once the first team's innings is completed the scoring rate of the second team is watched very keenly by players and spectators alike. Are they ahead of where their opponents were after the same number of overs? What will the effect be of the inevitable surge in scoring rate that occurs during the last few overs? The tactics are varied and if you have got a slow-scoring pair of batsmen at the wicket it might well be best for the bowling team to keep them there rather than get one of them out and risk a faster-scoring replacement.

The scoring-rate graph shows total runs scored versus number of overs bowled:

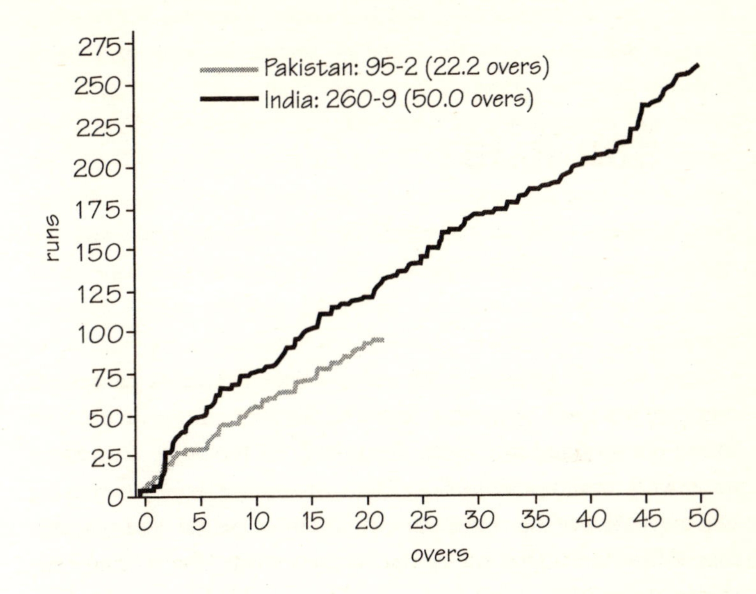

A number of things are interesting about the graph of each team's run-making progress. The height of the graph can only stay level or increase – you can't score negative runs! Although the running total is given by the vertical height of the graph, the scoring rate is given by the *slope* of the graph. The steeper the upward gradient, so the faster is the rate of scoring; and if no runs at all are scored off an over the curve will be horizontal.

If you try to read off the scoring rate from the graph you will notice a little problem. The rate depends upon how long a piece of the graph you use to estimate the slope. If you take the starting point (zero runs from zero overs) at the origin and the final score after 50 overs, let's say it is 250, then the run rate is just the total score divided by 50 and is 5 runs per over. Suppose that only 100 runs were scored in the first 25 overs, so the run rate would be only 4 runs per over after 25 overs and 6 per over during the second 25 overs. But if you look in finer detail at the graph you can see that it gets updated after each ball and so it has $50 \times 6 = 300$ steps along the horizontal axis at which it can be updated. The instantaneous run rate is just the slope of the curve between two of those points and can never be steeper than 6, the most runs you can hit off a single ball (excluding rare circumstances). The steepest overall gradient would be achieved in the bizarre situation where the batting team hit a six off every one of the 300 balls they received. Their score would be the maximum of 1,800 and the rate would be 36 runs per over. The highest international score achieved in 50 overs is 443 by Sri Lanka, just less than 9 runs per over.

The run-rate graph is made up of a linked staircase of straight lines. If we drew the most accurate possible smooth curve (no sharp corners and don't take the pencil off the page) through it then we would be able to determine the scoring rate by specifying the formula for the continuous curve S(N) for the score against overs, N. We know that S must be zero when $N = 0$ and if the final score is 250 after 50 overs then the curve might, for example, look like $S(T) = 250(N/50)_n$ where n is close to 1.

39

Squash – A Very Peculiar Practice

Some racket sports, like squash, and team sports like volleyball, have a history of using a distinctive scoring system in which you only score a point if you win it while serving. These games are played to a minimum of 9, 11 or 21 points, with a margin of 2 points required to win and then best of five games to win the match. The main drawback with this lugubrious system is that the length of games is extremely unpredictable. Points can be won and lost many times without there being any advance on the scoreboard. This is a real headache in scheduling matches in big tournaments, or if TV coverage is sought. The length of each match could easily differ by a factor of 2 from that of another with the same number of sets. As a result badminton switched from playing each game to 15 points[1] to a system of 21 with scoring occurring when you win a rally, whether or not you are serving, with the best of three games. In 2004, squash also changed to this type of PARS, or 'point a rally' scoring system, but playing to 11 points[2] with the match decided by the best of five games.

Before squash changed its scoring system, it employed another mathematically interesting rule. If the score reached 8–8, then the receiving player could choose whether the game should be played to 9 or to 10 points.

What should a player choose to do? If the receiver is on average the weaker player then it would be best to play to 9 but if the receiver is the stronger player then it is best to play to 10. The weaker player might luckily win 1 point but the chance of winning 2 in the same way is far more improbable.

If the probability of winning a point is p and R is the probability of *scoring* the next point from 8–8 when you are the receiver and S is the probability of *scoring* the next point if you are serving, then $R = pS$ because the receiver must first win a point to become the server. Working out S is straightforward. You can score by winning the next point because you are serving; the probability for this happening is p. But if you lose that next point (with probability $1 - p$) then your chance of scoring is just R because you now become the receiver. Therefore, we see that $S = p + (1 - p)R = p + (1 - p)pS$, and so:[3]

$$S = p/(1 - p + p^2) \text{ and } R = p^2/(1 - p + p^2)$$

Now we can decide whether the receiver should play to 9 or 10 from being tied at 8–8. If you play to 9 then your chance of winning is just the probability of scoring the next point as a receiver, which is R. If you choose to play to 10 then you can score the 2 points needed for victory by winning twice: first with a chance RS of winning the first as receiver, and then with chance S as server, to win via the score sequence 9–8, 10–8; or you could win by going 9–8, 9–9, 10–9 with probability $R(1 - S)R$, or by the sequence 8–9, 9–9, 10–9, which has probability $(1 - R)RS$. Adding these three probabilities together we get:

$$\text{Probability of winning if you play to 10} = RS + R^2(1 - S) + RS(1 - R)$$

Therefore, you are more likely to win by playing to 10 than to 9 if:

$$(1 - 2S)(1 - R) < 0$$

and so we must have $S > ½$.[4] This means that the probability of winning a point must satisfy $p/(1 - p + p^2) > \frac{1}{2}$ and this requires $p > \frac{1}{2}(3-\sqrt{5}) = 0.382$.[5]

If your chance of winning a point is greater than about 38% then play to 10. But if you know you are the weaker player with a less than 38% chance of winning each point, and so are fortunate to be hanging in the game at 8–8, then opt to play to 9. You might just fluke one more point but don't count on two.

40

Faking It

One of the most interesting general statistical misconceptions is what a random distribution is like. Suppose that you had to tell whether a particular sequence of events was random or not. You might hope to tell if it was not random, by finding a pattern or some other predictable feature. Let's try to invent some lists of 'heads' (H) and 'tails' (T) that are supposed to be the results of a sequence of tosses of a coin so that no one would be able to distinguish them from real coin tossings. Here are three possible fake sequences of 32 coin tosses:

THHTHTHTHTHTHTHTHTTTHTHTHTHTHTHH
THHTHTHTHHTHTHHHTTHHTHTTHHHTHTTT
HTHHTHTTTHTHTHTHHTHTTTHHTHTHTHTT

Do they look right? Would you regard them as likely to be real random sequences of heads and tails taken from true coin tosses, or are they merely poor fakes? For comparison, here are three more sequences to choose from:

THHHHTTTTHTTHHHHTTHTHHTTHTTHTHHH
HTTTTHHHTHTTHHHHTTTHTTTTHHTTTTTH
TTHTTHHTHTTTTTHTTHHTTHTTTTTTTTHH

If you asked the average person whether these second three sequences were real, most would probably say no. The first three sequences looked much more like their idea of being random: there is much more alternation and they don't have the long runs of heads and of tails displayed by the second trio. If you just used your computer keyboard to type a 'random' string of Hs and Ts you would tend to alternate a lot and avoid long strings otherwise it 'feels' like you are deliberately adding a correlated pattern.

Surprisingly, it is the second three sequences that are the results of a supposedly true random process. The first three, with their staccato patterns and absence of long runs of heads or tails, are the fakes that someone wrote down to fool you. We just don't think that random sequences can have all those long runs of heads or tails, but their presence is one of the acid tests for the genuineness of a random sequence. The coin-tossing process has no memory. The chance of a head or a tail from a fair toss is ½ each time, regardless of the outcome of the last toss. They are all independent events. Therefore the chance of a run of r heads or r tails coming up in sequence is just given by the multiplication $\frac{1}{2} \times \frac{1}{2} \times \frac{1}{2} \times \frac{1}{2} \times \ldots \times \frac{1}{2}$, r times. This is $\frac{1}{2}^r$. But if we toss our coin N times so that there are N different possible starting points for a run of heads or tails, our chance of a run of length r is increased to $N \times \frac{1}{2}^r$. A run of length r is going to become likely when $N \times \frac{1}{2}^r$ is roughly equal to 1, that is when $N = 2^r$. This has a very simple meaning. If you look at a list of about N random coin tosses then you expect to find runs of length r where $N = 2^r$. All our six sequences were of length $N = 32 = 2^5$ so if they are randomly generated we expect there is a good chance that they will contain a run of five heads or tails and they will almost surely contain runs of length four. For instance, with thirty-two tosses there are twenty-eight starting points which allow for a run of five heads or tails and on average two runs of each is quite likely. When the number of tosses gets large we can forget about the difference between the number of tosses and the number of starting points and use $N = 2^r$ as the handy rule of thumb. The absence of these runs of heads or tails is what should make you suspicious about the first three sequences and happy about the likely randomness of the second three. The lesson we learn is that our intuitions about randomness are biased towards thinking it is a good deal more ordered than it really is. This bias is manifested by our expectation that extremes, like long runs of the same outcome, should not occur – that somehow those runs are orderly because they are creating the same outcome each time.

These results are also interesting to bear in mind when you look at long runs of sporting results. One famous losing streak that has a random origin is when the England cricket captain, Nasser Hussain, lost the toss at the start of all the fourteen international matches in which he was captain during 2000–1. Strikingly, he first lost seven in a row, then missed a match in which his replacement won the toss, before returning to lose it on a further seven consecutive occasions. He had only a 1 in 16,384 ($= 2^{14}$) chance of losing fourteen separate coin tosses. However, since he captained England about a hundred times the chance of a fourteen-game losing streak is a hundred times smaller, or about 1 in 164. This is still pretty improbable but more plausibly down to very bad luck.

41

A Sense of Proportion

As you get bigger, you get stronger. We see all sorts of examples of the growth of strength with size in the world around us. A small kitten can hold its spiky little tail bolt upright yet its much bigger mother cannot: her tail bends over under its own weight, demonstrating that strength does not increase in direct proportion to volume. The superior strength of heavier boxers, wrestlers and weightlifters is acknowledged by the need to grade competitions by the weight of the participants. But how fast does strength grow with increasing weight or size?

Simple examples can be very illuminating. Take a short breadstick and snap it in half. Now do the same with a much longer one. If you grasped it at the same distance from the snapping point each time you will find that it is no harder to break the long stick than to break the short one. A little reflection shows why this should be so. The stick breaks along a slice through the breadstick. All the action happens there: a thin sheet of molecular bonds in the breadstick is broken and it snaps. The rest of the breadstick is irrelevant; if it was a hundred metres long it wouldn't make it any harder to break. The strength of the breadstick is given by the number of molecular bonds that have to be broken across its cross-sectional area. The bigger that area, the more bonds need to be broken, and the stronger the breadstick. So strength is proportional to the cross-sectional area, which is proportional to some measure of its diameter squared.

Breadsticks and weightlifters have a constant density that is just determined by the average density of the atoms that compose them. But density is proportional to mass divided by volume, which is proportional to size cubed. Sitting here on the earth's surface mass is proportional to weight and so we expect the simple proportionality 'law' that for a fairly spherical object:

$$\text{Strength} \propto (\text{weight})^{2/3}$$

This simple rule of thumb allows us to understand all sorts of things. The ratio of strength to weight is seen to fall as strength/weight $\propto$ (weight)$^{-1/3}$ $\propto$ 1/(size). So as you grow bigger, your strength does not keep pace with your increasing weight. If all your dimensions expanded uniformly in size, you would eventually be too heavy for your bones and you would break. This is why there is a maximum size for structures made of atoms and molecules on the earth's surface, whether they are dinosaurs, trees or buildings. Scale them up in shape and size and eventually they will grow so big that their weight is sufficient to break the molecular bonds at their base and they will collapse under their own weight.

We started by mentioning some sports events where the advantage of size and weight is so dramatic that competitors are divided into different classes according to their body weight. Our 'law' predicts that we should expect to find a straight line when we plot the cube of the weight lifted against the square of the body weight of weightlifters.

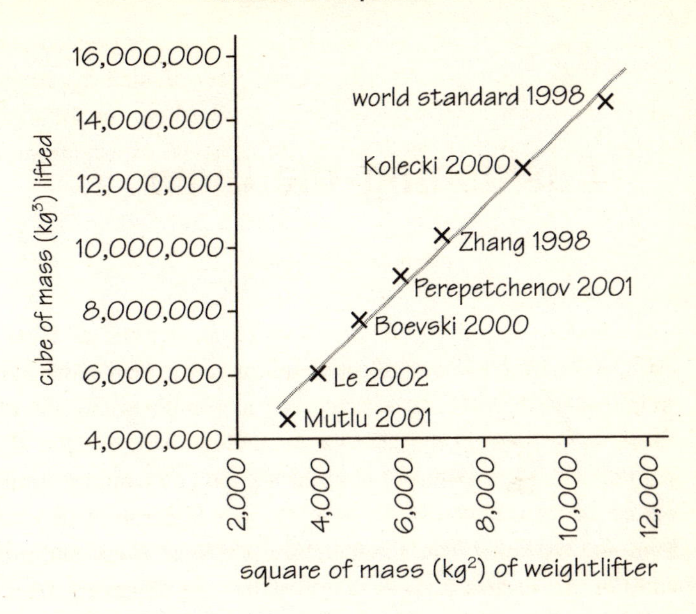

Here is what happens when you plot that graph for the recent men's world records in the clean-and-jerk across the weight categories. Sometimes mathematics can make life simple. The weightlifter who lies farthest above the line giving our rule (*) is the strongest lifter, 'pound for pound', whereas the heaviest lifter, who lifts the largest weight, is almost the weakest when his weight is taken into account.*

*From 1896 to 1932, rope climbing was an Olympic event. Competitors had to be the quickest to reach the top of a fourteen-metre-high rope. This event is a straight battle between strength and weight and the lighter the competitor the better.

42

Cushioning the Blow

Most of us are familiar with the existence of a 'sweet spot' on a tennis racket or a bat: the place where it is best to strike the ball. What 'best' means is that there is no reaction back on your hand at the place where you are holding the bat. This spot is located where the motion of the bat as a whole in response to the force from the incoming ball is equal and opposite in direction to the rotation of the bat about its centre of gravity. Physicists call this sweet spot the 'centre of percussion' and it occurs at a distance from the top of a cricket or baseball bat that is about two thirds of its total length.

Snooker and pool balls have sweet spots too. If you strike a ball with the cue parallel to the table then where you strike it determines how it will move. The point of contact with the table acts like the pivotal point where you grip a bat. Obviously, if you aim your shot through the centre of the ball then the ball will slide as a whole across the table without any rolling. Strike it a little higher than the centre and it will both slide and rotate. The sweet spot for the ball is the point on the ball where the speed away from the cue caused by the sliding is equal to the rotational velocity in the opposite direction caused by its rolling. Hit the ball at the height of this sweet spot and it will not slide; it will start rolling straight away. Where is this special spot?

The height of the sweet spot above the table is:

$$h = r + I/Mr$$

where r, M, and $I = 2Mr^2/5$ are the radius, mass and moment of inertia of the ball, respectively. So, we see that $h = 7/10 \times 2r$; that is, the sweet spot is at a height equal to 0.7 times the diameter (2r) of the ball.

I thought that this would also tell me the height of the cushion around the sides of a pool or snooker table. My reasoning was that you should design a table for a cue ball game so that the rebounds of the balls from the sides of the table are as smooth and true as possible. If the height of the point of impact on the cushion was 0.7 times the ball diameter, it would mean that after a ball hit the cushion it wouldn't slide across the table, just roll, losing very little energy in the process. I went to look up the rules governing the game. Alas, I was slightly disappointed. The official snooker specification is that the cushion height must be 0.635 ± 0.01 times the ball's diameter. Almost right, but not quite, it seems. I was puzzled, so I asked snooker and pool dynamics expert David Alciatore why 0.635 isn't 0.7. His answer injected a little reality into the issue. It seems that the height is slightly lower than the sweet spot height of 0.7 times the ball diameter so as to reduce the downward push of the rebounding ball on to the surface of the table near the cushion (called the 'gutter') and reduce wear on the table surface.

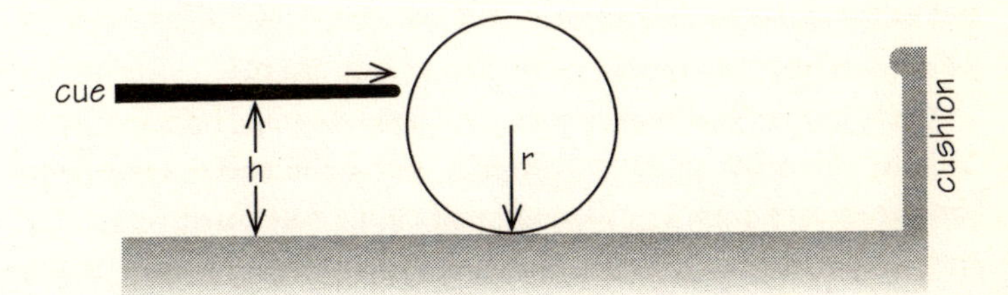

43

Breaststrokers

The familiar racing swimming strokes are, in descending order of speed, front crawl, butterfly, backstroke, and breaststroke.[1] At present, the world records for men over 50m are 20.91s (crawl), 22.43s (butterfly), 24.04s (backstroke) and 26.67s (breaststroke) while the records for women are 23.73s (crawl), 25.07s (butterfly), 27.06s (backstroke) and 29.80s (breaststroke). The gap between men and women is consistently about three seconds over this distance, regardless of stroke.

Front crawl is not a prescribed stroke but in practice it is the fastest way to move through the water and so is always used in competitions where swimmers are free to use any stroke style they wish.[2] There are other swimming strokes like sidestroke, or trudgeon (crawl arms and breaststroke legs), inverted breaststroke (breaststroke on your back), slow butterfly (using breaststroke legs) and other combinations of arm and leg strokes, but they are used for recreational swimming or lifesaving purposes.

There are certain symmetries that define the four competition strokes. Butterfly requires complete left–right body symmetry: both arms and both legs move in pairs in unison. Backstroke and crawl both display an asymmetric motion with alternate right and left arm pulls on the water while the leg movements are up and down with any pattern. In the crawl there is a breathing choice, either every stroke or every two, three or four strokes. Top swimmers tend to breathe on alternate sides as it helps to keep right–left body symmetry through the water. As a result, each of these three strokes results in the swimmer being propelled forwards (except at a turn) at a fairly constant speed. At any moment, either the right hand or the left hand applies a force to pull water backwards and the body is forced forwards in equal and opposite reaction to this, just as Isaac Newton taught us. These three strokes also share the common feature that the recovery phase of the arm motion takes place out of the water.

The fourth stroke, the oldest established and slowest, is the odd one out. Breaststroke also has lateral symmetry but all the arm and leg movements must take place approximately in the horizontal plane parallel to the water surface, rather than perpendicular to it as in the other three strokes, although there is significant bobbing up and down in practice. There are strict rules about the symmetry of the arms and legs, the elbows and legs must always be kept under the water and part of the head must always be above the surface after the start or turn is completed for the stroke to be legal. As this is the slowest stroke, any illegal modifications can be very advantageous. At the 1956 Games in Melbourne, several breaststroke swimmers swam long sections of the race completely submerged. This strategy reduced the water resistance significantly because it removed the friction created by breaking the surface continually. The Japanese swimmer Masaru Furukawa competed in accord with the rules at the time by swimming all of the first three lengths and most of the final length underwater to win the 200m event. Subsequently, the rules were changed to prevent this trick – not least because swimmers were passing out because of severe oxygen deficiency! Over the past twenty-five years, small variants of the arm action have evolved and a single dolphin kick is permitted away from each turn.

The breaststroke is also unique in that swimmers do not move through the water at a roughly constant speed. The drag of the water is always trying to decelerate the swimmer. This force is considerable because it is proportional to the swimmer's speed – the faster they go the greater it becomes. The propulsive force they generate by moving water backwards during the first cycle of their stroke is accelerating them, but it is countered by a decelerating force that arises as they move water forwards by bringing their knees up and their arms forwards ready for the next stroke.[3]

The breaststroker's speed is at a maximum when the arms are extended forwards and the legs are straight. It is at a minimum, and close to zero, when the arms are fully extended sideways and the knees have been brought right up towards the body. At that moment, the water drag and the forward motion of the water combine to produce the maximum deceleration. As the arms pull the water backwards, and the legs push back, the motion is accelerated again and the speed builds up. The swimmer's speed is therefore always changing between almost zero and about 2m/s as a result of a periodic variation of the net force on his body from the three forces acting on him. He never stops, of course, because even with zero net force acting momentarily there is still residual momentum. A graph of these variations is shown on the next page. Breaststroke is a complicated piece of staccato human movement! The diagram on the next page shows the variation in time of the swimmer's speed and acceleration through the entire breaststroke cycle.

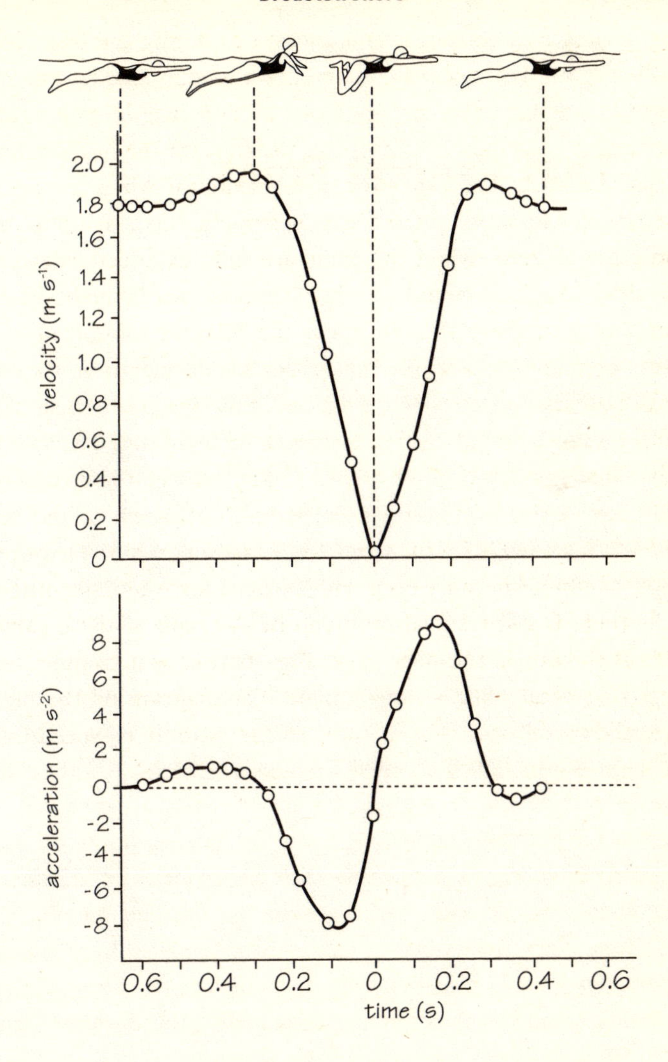
2.0
1.8
1.6
1.4
1.2
1.0
0.8
0.6
0.4
0.2
0
velocity (m s⁻¹)
8
6
4
2
0
-2
-4
-6
-8
acceleration (m s⁻²)
0.6 0.4 0.2 0 0.2 0.4 0.6
time (s)

44

That Crucial Point

How many times have you heard tennis commentators tell you that a particular point in the game is the most important one to win? Curiously, there was a traditional view, supposedly originating with Pancho Gonzales, that when a player was serving the crucial point was at 15–30. This is clearly not true. If the player loses the next point he can always recover from being 15–40 down. But if the score is 15–40, or 30–40, or advantage to his opponent (which is actually the same thing as 30–40) and he loses the next point then he has lost the game.

Tennis has an unusual scoring system. It could just keep a single cumulative score like table tennis – first to score at least 11 or 21 points with a margin of at least 2 points. Then best of three sets. Instead we have games and sets with points being accumulated in 15, 30, 40, perhaps originating from the quarter-hours on the courtside clock when the game was first played in medieval France. The winner would have had his score go through the 15, 30, 45-minute clock points before victory was signalled by reaching the 60. Supposedly, the shift from 45 to 40 was to allow the use of the 40, 50 and 60 slots to cover advantage (move forward from 40 to 50) and deuce (move back from 50 to 40).

Many of the archaic terms in the game have French origins: deuce from the French '*deux*', signalling that 2 more points need to be won; 'love' for zero is derived from '*l'oeuf*' for 'egg', the symbol of a zero – a 'goose egg' is still a term for zero in American sport. Even the name of the game is thought to derive from the

French '*tenez*' ('take that') that was shouted at the opponent as a serve was unleashed.

The scoring system has the property that it maintains interest in the game, from players and spectators, over a much longer period of time than would be the case if the scoring ran up to a large number like 21. Even though a player loses a set 6–0, all is square again for the start of the next set and things are much more competitive than if the weak player had to carry on with the 6–0 millstone around his (racket) head. Sometimes matches are played so that the best of three sets wins and sometimes (for men only) the best of five wins. If players are very closely matched then you need more sets to ensure that the better player wins in the long run and doesn't succumb to flukes so easily. It seems that it is still thought that the men's matches are closer-fought affairs than the women's because the women only play three sets (yet insist on the same prize money as the men). There doesn't seem to be a good reason for this inequality. The women players are fit enough to play five sets, especially as tiebreaks allow only the final set to run its potentially lengthy course.

45

Throwin' in the Wind

In the Olympic throwing events the effects of wind in the stadium are subtle. The shot and the hammer are so heavy that wind makes no significant difference to their trajectories but the javelin and the discus are aerodynamic and their flight is significantly affected by the wind speed and direction. Part of the skill of throwing these objects is an ability to exploit the wind in an optimal way.

The javelin thrower has a run-up of about 30–35m and must launch the javelin without allowing any part of the body to touch the ground on or beyond the curved foul line. The javelin must make a mark when it lands on the grass, and must fall within the safety sector defined by a fan opening at a 29-degree angle that radiates out from the curved throwing line, so a top-class thrower has a 'target' span of about 45m to aim at for a valid throw. Only about a quarter of the javelin's launch speed relative to the ground comes from the thrower's run-up; the rest is due to the speed of the launch arm. Past Olympic champions like Janis Lusis used only three or four steps before they threw. Top javelin throwers are the people who could throw cricket balls the farthest in school: they have remarkable arm speed rather than the brute strength of the shot-putter. However, the strain imposed on the arm and shoulder muscles by the sudden wrench of throwing is huge and a thrower's career is often ended prematurely by arm or shoulder injuries.

The men's javelin weighs only 800gm (the women's is 600gm) and is much lighter than the shot, discus or hammer – normal-sized athletes have a chance here. The greatest ever javelin thrower is the Czech Jan Zelezny, who is now an IOC member. He won Olympic silver in 1988, gold in 1992, 1996 and 2000, but weighs only 88kg and is 1.86m tall. His world record stands at 98.48m.

When a javelin is launched directly into a significant wind it is better to go for a slightly flatter trajectory of around 35 degrees (i.e. a smaller launch angle than the optimal 44–45 degrees in still air). It then hangs in the air for longer, and sails out to a greater distance. A good thrower will ensure that the javelin is angled to lie at about 10 degrees – the so-called 'angle of attack' – below the trajectory followed by the centre of gravity of the javelin, as shown in the diagram here. This ensures that the nose of the javelin is always dropping and is inclined to penetrate the ground when it lands.

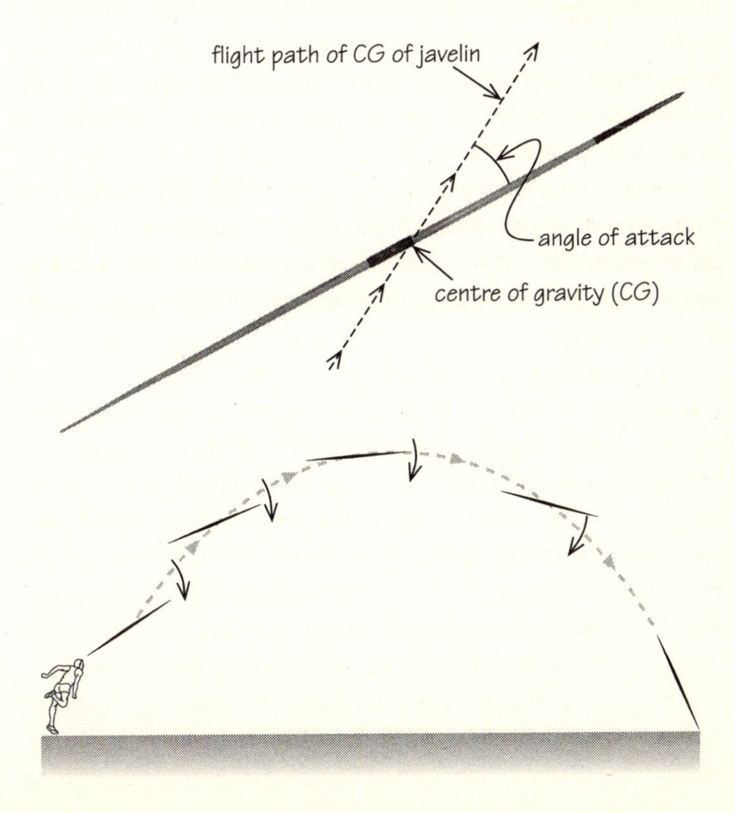

A key factor in determining the trajectory of the javelin is the location of its centre of gravity. Back in 1984 the East German athlete, Uwe Hohn, threw the javelin an astonishing 104.8m to create a new world record. This was too dangerous for events in an arena with judges and athletes milling around in other competitions, especially if the javelin landed flat and skimmed along the grass at high speed like a runaway torpedo. The business of whether a javelin had made a mark or fallen flat was also a pretty contentious matter for judges and competitors to agree on. Hohn was likely to clear the grass completely and end up with his javelin hitting the track or the high-jump fan. The response in 1986 was to alter the javelin, moving its centre of gravity forward by 4cm and redesigning the tail to make it less aerodynamic. The smaller, women's javelin was similarly redesigned in 1999. As a result, the new javelins didn't hang flat in the air for so long: they nose-dived and invariably hit the ground at a steep angle and stuck in the grass. Hohn's great record was consigned to the history books and the event had a new start. Yet although the redesign reduced the distances thrown by about 10%, Zelezny had thrown the new javelin more than 98m by 1996 and safety problems were looming again – but his record has not been approached for more than fifteen years.

46

The Two-headed League

In 1981, the English Football Association made a radical change to the way its leagues operated, in an attempt to reward more attacking play. They proposed that 3 points be awarded for a win, rather than the 2 points that had traditionally been the victors' reward. A draw still received just 1 point. Soon, other countries followed suit and this is now the universal system of point scoring in football league competitions. It is interesting to look at the effect this has had upon the degree of success that an average non-winning team can have. When 2 points were awarded for a win it was easily possible to win the league with 60 points from forty-two games and so a team that gained 42 points from drawing all its games could finish in the top half of the league – indeed, Chelsea won the old First Division championship in 1955 with the lowest ever points total of 52. Today, with 3 points for a win, the champion side needs over 90 points from its thirty-eight games and an all-drawing side will find its 38 points will put them 3 or 4 from the bottom, fighting against relegation.

With these changes in mind, let's imagine a league where the football authorities decide to change the scoring system just after the final whistles blow on the last day of the season. Throughout the season they have been playing 2 points for a win and 1 point for a draw. There are thirteen teams in the league and they play each other once, so every team plays twelve games. The All Stars win five of their games and lose seven. Remarkably, every other game played in this league is drawn. The All Stars therefore score a total of 10 points. All the other teams score 11 points from their eleven drawn games and seven of them score another 2 points when they beat the All Stars, while five of them lose to the All Stars and score no more points. So seven of the other teams end up with 13 points and five of them end up with 11 points. All of them have scored more than the All Stars who therefore find themselves languishing at the bottom of the league table.

Just as the despondent All Stars have got back to the dressing room after their final game and realise they are bottom of the league, facing certain relegation and probable financial ruin, the news filters through that the league authorities have just voted to introduce a new points scoring system and apply it retrospectively to all the matches played in the league that season. In order to reward attacking play they will award 3 points for a win and 1 for a draw. The All Stars quickly do a little recalculating. They now get 15 points from their five wins. The other teams get 11 points from their eleven drawn games still. But now the seven that beat the All Stars only get another 3 points, while the five that lost to them get nothing. Either way, all the other teams score only 11 points or 14 points and the All Stars are now the champions!

47

What a Racket

Some things are harder to move than others. Most people think that the only problem is their weight: the heavier the load, the harder it is to shift it. But try moving lots of different types of load and you will soon discover that the distribution of the weight plays a significant role. The more concentrated the mass is towards the centre, the easier it is to move the object and the faster it rolls. Look at an ice skater beginning a spin. They will start with their arms outwards and then steadily draw them in towards their body. This results in an increasingly rapid spin rate. As the skater's mass is more concentrated towards the centre, they move faster. On the other hand, girders used to build robust buildings have an H-shaped cross-section, which distributes more mass away from the centre and makes it harder to move or deform the girder when it is stressed.

This resistance to being moved is called 'inertia', following its common usage, and it is determined by the total mass of an object and also by its distribution, which will be determined by the shape of the object. If we think about rotating an object then a simple tennis racket is an interesting example. It has an unusual shape and can be rotated in three distinct ways. You can lay the tennis racket flat on the floor and spin it around its centre; you can stand it on its top and twist the handle; and you can hold it by the racket handle and throw it up in the air so it somersaults and returns to be caught by the handle again. There are three ways to spin it because there are three directions to space, each at right angles to the other two, and the racket can be spun around an axis pointing along any one of them. The racket behaves rather differently when spun around each of the different axes because its mass distribution is different around each axis and so it has a different inertia for motion around each axis.

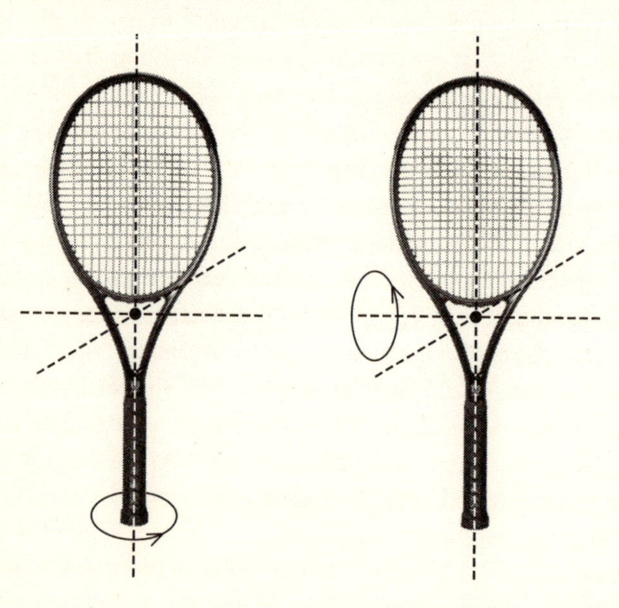

There is one remarkable property of these different motions which is displayed by tossing the racket. The motion around the axes about which the inertia is the largest or the smallest is simple. When the racket is flat on the ground or stood upright and spun like a top it does nothing very unusual. But when you spin it about the *in-between* axis, about which the inertia is intermediate between the largest and the smallest, something unusual happens. Hold the racket by the handle with the head horizontal as if you were holding a frying pan. Mark the top face with some chalk. Toss the racket so that it does a complete 360-degree turnover and catch it by the handle. The side with the chalk mark will now be facing downwards.

The golden rule is that a spin around the axis with the in-between inertia is unstable. The slightest deviation from the precise centre line always causes it to flip over. Sometimes this is a good thing. If you are a gymnast doing a somersault on the beam then you look more impressive (and score higher marks) if you do a twist as well. But the twist happens automatically because of this instability if you arrange your body right. If you are a high-board diver wanting to do a sequence of fast somersaults without any twists then you tuck up your body so that you spin about the axis with the smallest inertia and no twist happens.

48

Size Matters

There is something distinctly odd about many sports that place a premium on strength. As we have already seen, some pay great attention to the weight and size of the competitors and confine them to weight categories so that they only compete against opponents with fairly similar weights. The most obvious examples are in weightlifting and the combat sports like boxing, wrestling and judo. Even in non-Olympic rowing, there are separate races for 'lightweight' crews. The size differences between leading competitors in different weight classes are vast, and competitors will need to control their weight quite carefully if they are not inadvertently to fail to 'make the weight'.

The reasons for these segregations by weight are clear. The bigger you are so the stronger you should be. We have already seen that a person's strength increases like the two-thirds power of their weight, so if you cube your weight it will square your strength. This simple rule was borne out quite well when we looked at the trends in the world weightlifting records.

All this seems entirely fair and reasonable and we would expect only to find that the actual body weights of lifters and boxers will crowd towards the upper limit allowed for their weight class:* they want to make the most of the strength they can create by increased muscle weight. Yet when we look at the field events in the Olympic

*At the 1904 Olympics the American boxer Oliver won gold medals in both the bantamweight and featherweight divisions, the only time this feat has been achieved by a boxer.

athletics programme there seems to be a puzzling inconsistency. There are no weight classes for shot-putters, or discus and hammer throwers, where strength is a key factor in determining performance.* One consequence of this is that these throwers are all huge. There is no point aspiring to be a shot-putter if you have the build of a flyweight boxer or a figure skater. The increase of strength with body weight (and hence of muscle weight) ensures that lightweights avoid these events and encourages heavyweights to get heavier by building muscle through weight training and diet.

It is important to appreciate that throwers are by no means lumbering hulks. They are tremendously dynamic and can generate extraordinary speed in the throwing circle before launching their throwing implement. Many years ago there was a famous challenge sprint race over 200m between the shot-putter (and later 'the world's strongest man') Geoff Capes and distance runner Brendan Foster at the Crystal Palace stadium during an international invitation meeting. It seemed to have been inspired by some ill-judged remarks by TV commentators about Capes, who weighed 23 stone and was 6' 7" tall. At the time Foster was the world record holder for 3,000m and had won many medals at 1,500, 5,000 and 10,000m. Most people would have betted on Foster winning a sprint race against Capes by a wide margin. To their great surprise (and his) Foster was left far behind as Capes winged away to finish in a very brisk 23.7s. Sprinting is about strength too.

It is simple to estimate how the effect of body weight should translate into longer throwing capability. In an event like the hammer the competitor uses his or her strength to swing the hammer in a circle. Through a series of rotations the hammer is

*Back in the Stockholm Olympics in 1912, there were two-handed throwing events for the shot, discus and javelin. Each competitor would throw with their left hand and their right hand and the winner was the competitor with the greatest combined distance.

accelerated and released at an optimal angle within a prescribed sector of the field, after which it flies as a projectile. The athlete has got to avoid stepping outside the throwing circle. The stronger the athlete, the higher the hammer rotation speed that he can support, and so the greater the launch speed of the hammer. If we ignore the effects of air resistance, the range of a projectile is proportional to the square of the launch speed, and if the radius of the circle in which the hammer turns is kept constant then the square of the speed at which it moves in a circle is proportional to the strength of the thrower (measured for example by the weight they can lift). Since we have seen that strength increases as the two-thirds power of the thrower's weight, we conclude that the distance thrown will increase accordingly – other things being equal.[1] Clearly, size matters – quite a lot.

49

A Truly Weird Football Match

What is the most bizarre football match ever played? In that competition I think there can only be one winner. It has to be the infamous encounter between Grenada and Barbados in the 1994 Shell Caribbean Cup. This tournament had a group stage before the final knockout matches. In the last of the group stage games Barbados needed to beat Grenada by at least two clear goals in order to qualify for the next stage. If they failed to do that, Grenada would qualify instead. This sounds very straightforward. What could possibly go wrong?

Alas, the law of unforeseen consequences struck with a vengeance. The tournament organisers had introduced a new rule in order to give a fairer goal difference advantage to teams that won in extra time by scoring a 'golden goal'. Since the golden goal ended the match you could never win by more than one goal in such a circumstance, which seemed unfair. The organisers therefore decided that a golden goal would count as two goals. But look what happened.

Barbados soon took a 2–0 lead and looked to be coasting through to the next phase. Just seven minutes from full time Grenada pulled a goal back to make it 2–1. Barbados could still qualify by scoring a third goal but that wasn't so easy with only a few minutes left. Better to attack your own goal and score an equaliser for Grenada because you then had the chance to win by a golden goal in extra time, which would count as two goals and so Barbados would qualify at Grenada's expense. Barbados obliged by putting the ball into their own net to make it 2–2 with three minutes left. Grenada realised that if they could score another goal (at either end!) they would go through, so they attacked their own goal to score that losing goal that would send them through on goal difference. But Barbados resolutely defended the Grenada goal to stop them scoring, and the match went into extra time. The Barbadians now took their opponents by surprise and scored the winning golden goal in the first five minutes. If you don't believe me watch it on YouTube.[1] It's a fitting tribute to FIFA.

50

Twisting and Turning

When wheels turn what determines how fast they spin? For the racing cyclist on the road or zooming around the velodrome this is a crucial question. Can you gain a marginal advantage by engineering a more efficient wheel? What is the key consideration?

When a body is tugged along its edge in order to make it spin it is not simply its mass that determines how fast it is going to spin about its axis, it is its mass distribution. We have already seen that the combination of mass and its distribution within the body determines its 'inertia' – and how hard it is to move it.

The inertia of a body is given by cMR^2 where M is its mass, R its radius, and c is a number that depends on how concentrated the mass distribution is around its centre. For a sphere of uniform density $c = 2/5$, but for a hollow spherical shell of the same size with all the mass at the surface, $c = 2/3$. So, c gets larger as the mass gets distributed farther from the centre. A hollow shell has a higher inertia, and rolls slower, than a solid sphere of the same mass and radius. The same principle holds when we compare the inertia of a hollow circular ring (MR^2) with that of a uniform circular disk ($\frac{1}{2} MR^2$). These last two quantities give good approximations to the inertia of different bicycle wheels. My everyday bike has wheels comprised of a ring with lightweight radial spokes for strength. This has a relatively high inertia, and so will respond to an application of force to the pedal more slowly than a wheel that is designed as a solid disc. This is why disc wheels are seen on top-flight racing bikes. Their inertia is low and they move quicker when force is applied by the chain by pushing on the pedals. However, a disc wheel on the front of your bike is horribly impractical unless you are always going to move in a perfect straight line, or perpendicular to the velodrome track surface, because the slightest deviation of the wheel to the side will catch the passing air, twist the wheel around, and pitch you on the ground. This is why you see disc wheels only as fixed rear wheels on competition bikes for road races.

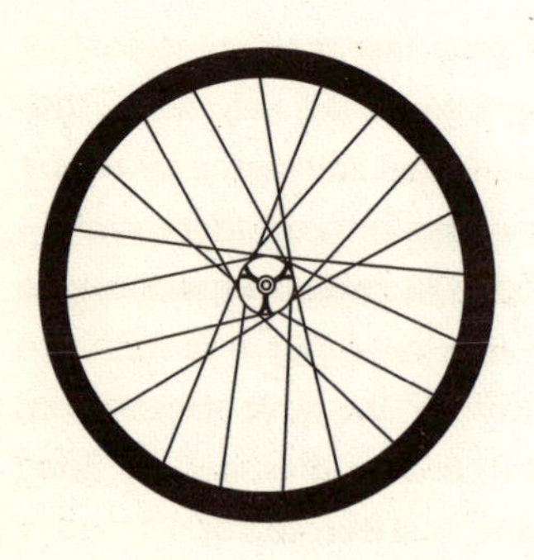

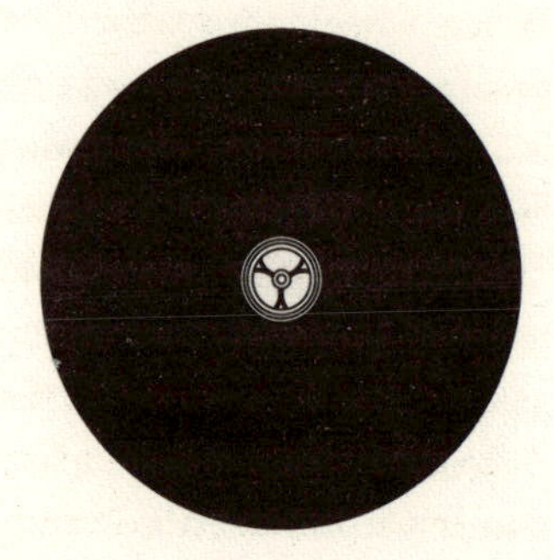

51

The Wayward Wind

Whereas athletics spectators worry about whether it is raining or cold, the prime concern of competitors is usually whether or not it is windy. Sprinters might welcome a tailwind so long as it isn't too strong to set a valid record, but anyone whose race involves one or more laps of the track hates wind. It slows times and makes you work harder. But some of the implications of wind are not as obvious as you look at different events in the athletics stadium.

In the 100m and 200m sprints, the 110m hurdles, and the long and triple jumps, a criterion for 'wind assistance' exists and a performance will not be valid for record purposes if the following wind speed exceeds 2m/s.[1] A following wind will be denoted with a + sign on the results and a headwind by a – sign. However, although a +5m/s wind will stop your 9m long jump from becoming the world record, it does not affect its validity in the competition. It still counts and if the wind then drops to zero for all the jumps taken by your fellow competitors then that is just too bad for them! This is particularly awkward in the qualifying rounds of the sprint races which involve many races over a long period of time. The qualifying conditions to get to the next round from eight heats will generally be something like 'the first three finishers plus the next eight fastest losers overall who didn't place in the first three'. If the wind speed is changing from race to race then this can be quite unfair on runners outside the first three who didn't benefit from a stronger following wind. Typically, the advantage gained from a 2m/s following wind over the 100m is about 0.1s compared with running in windless conditions, but running into a –2m/s headwind will slow you by slightly more than 0.1s.[2]

In the 200m, the direction of the wind is an even more significant factor because half of the race is run in lanes and runners in different lanes can feel more or less of a headwind in their faces or a tailwind behind their backs for different periods of time when they sprint around the curve because they start at different points around the staggered lanes.

In the 110m hurdles a new consideration comes into play. Although a following wind makes you run faster, the effect can be undesirable. High hurdlers have a very precise pattern of strides imprinted into their brains and nervous systems by countless training sessions and a following wind will make you arrive closer to hurdles than your stride pattern planned for. When a strong tailwind blows, you will see lots of hurdles fall as the hurdlers get blown too close to the hurdles as they take off. In the past, you were judged ineligible to set an Olympic record if you knocked over a hurdle (although, perversely, you could remain the winner of the race) and in 1924 an Olympic record was set by the third-placed athlete after the winner knocked down a hurdle and the second placer used an improper hurdling action.

Something similar happens in the long and triple jumps. Jumpers mark out their approach runs very precisely so as to place their take-off foot on the firm wooden take-off board as close as possible to the plasticine line that divides a legal jump from a foul. With a strong following wind your approach to the board will be significantly faster and you are much more likely to overstep the line. But get it right by adjusting your start checkmark to allow for this and the following wind will lead to a greater take-off speed and a longer valid jump. With a headwind, you will take off slower and be more likely to do so well behind the take-off board. You will lose distance on your jump from the lower take-off speed as well as the earlier take-off point.

Pole-vaulters benefit from a following wind on their approach run and this makes a big difference to their take-off energy and their clearance height because it depends on the square of the take-off speed (just like the long jump); yet there is no wind-assistance criterion for pole-vault performances to be legal for record purposes.

Running whole laps around a track in the wind is always harder work than running at the same speed in windless conditions. Imagine a square track with wind of speed W blowing in a direction parallel to one of its sides. The power (= force × velocity) needed to run a lap around its four sides (one with a tailwind, one with a headwind, and two with neutral crosswinds* as shown in the figure below) at speed V is proportional to $2V(V^2 + W^2) + V(V - W)^2 + V(V + W)^2 = 4V^3 + 4VW^2$. You notice that the total power factor ($4V^3$) needed when there is no wind is always less than when there is a non-zero wind speed W or –W. The tailwind gain never makes up for the effect of the headwind and the crosswinds. The crosswind is always adverse. The optimal strategy is to shelter behind other runners when there is a headwind and ensure none are directly behind you in a tailwind by changing lanes.

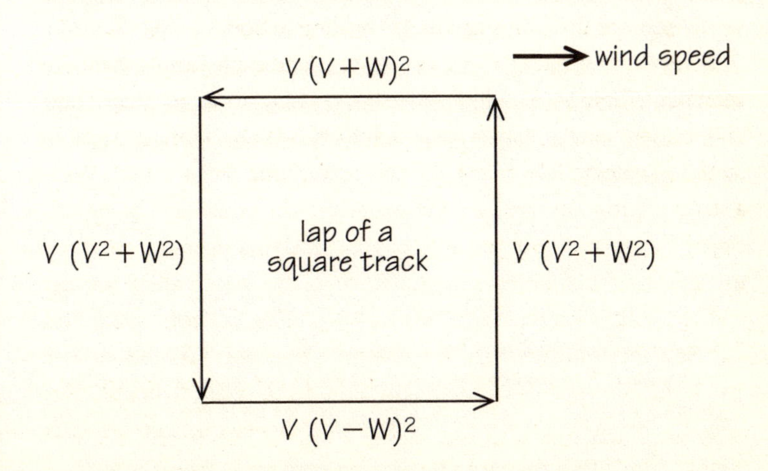

The last and oddest thing to say about wind assistance in the track events concerns the accuracy of its measurement. Athletes' times are recorded to hundredths of a second for record purposes and extraordinary trouble is taken to get them right and to determine the reaction times of athletes electronically to hundredths of a second so as to rule out false starts. (Incidentally in the ancient Greek Games, runners who false started were whipped by the officials!) By contrast, the accuracy of wind readings is nowhere near as good. The recorded wind speeds need to be reliable to an accuracy of at least 0.2m/s to reflect the accuracy of timings for record purposes. Studies have found that the official method of recording wind speed at the 50m point in the 100m straight with one IAAF-approved tube-propeller anemometer was only likely to be accurate to about ± 0.9m/s.[3] This is like working with race timings that were only accurate to 0.05s. An experimental study of wind gauge use by Nick Linthorne found that the wind speed recorded can change significantly during the ten-second duration of a 100m race.

By placing wind gauges at different points along the 100m straight, significant differences in wind speed were found, and this is the source of the ± 0.9m/s uncertainty reported. The situation in the 200m was not studied in the same way but it would undoubtedly turn out to be even more complicated because of the different effective winds felt by athletes in different lanes. An athlete in lane one might run a 'legal' time whilst a runner in lane eight is 'wind-assisted'!

* In the crosswind you always feel a resultant wind speed of $V_r = (V^2 + W^2)^{1/2}$. The drag force, F, is proportional to V_r^2 and the power required to overcome it is propotional to $FV = V(V^2 + W^2)$. We ignore the small extra drag you will experience if your body cross-sectional area is larger and less aerodynamic when impacted partly from the side by the air at an angle whose tangent is W/V.

52

Windsurfing

Windsurfing looks tiring. It can be enticing in the warm Caribbean sea, but off the British coast the water temperature is a tad lower and the sea is not quite so crystal clear. Competitive windsurfing has been part of the Olympic Games since 1984, but has struggled to remain in the face of falling popularity brought about by disputes about the commercialisation (and spiralling cost) of equipment in the 1990s.

Windsurfers negotiate a trapezoid-shaped circuit, like a large running track, and score points during a sequence of races each day, lasting thirty to forty-five minutes. This is a strenuous endurance event and training is a full-time job. The windsurfer has to remain balanced on a surf board, two to four metres in length, by countering the pull on the boom connected to the mast. This requires both strength and balance. We are going to work out just how much strength is required.

The set-up of the surfer balanced on the board is shown in the figure below. This event is the geometric mean of surfing and sailing!

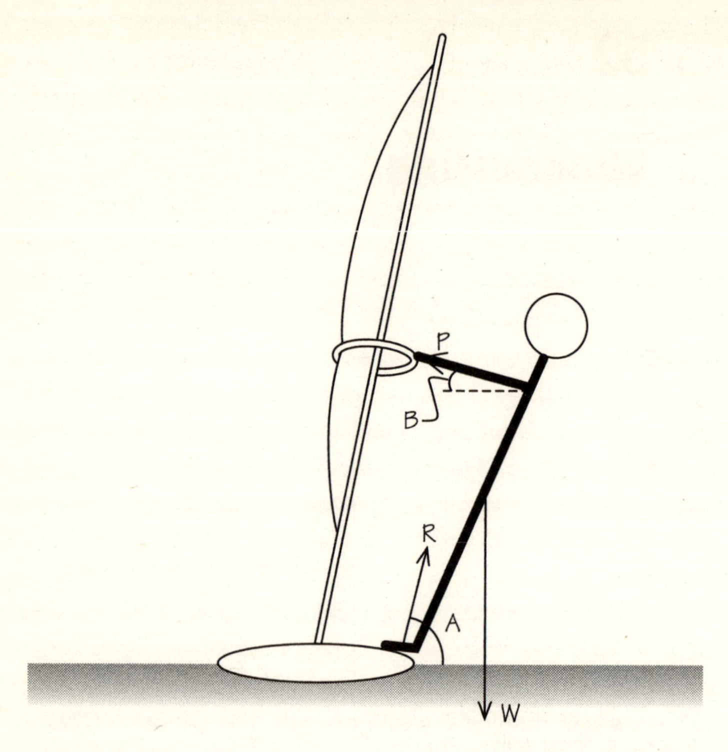

We are going to neglect the mass of the mast and sail because it is so much less than that of the windsurfer. The total mass of the boom, the mast and the sail is typically about 10kg, but the weight of the surfer is about 65–70kg. There are three dominant forces acting here: the weight of the windsurfer, W = Mg; the reaction force of the board on his feet, R; and the pull on the surfer's arms, P. If we require these forces to be in balance in the vertical direction then Mg = RsinA + PsinB. And for the horizontal forces to balance we also need RcosA = PcosB. Combining these we can calculate the pull that must be exerted by the surfer's arms in terms of his weight:

$$P = Mg/(\sin B + \cos B \tan A) \quad (*)$$

The last criterion we need to apply for stability is that the moments of the forces acting about, say, the surfer's feet must sum to zero, otherwise there would be an overall torque that would quickly lever him into the water. We need the moments (force × distance) of his weight and the pull on the arms. The reaction force acts at the point we are taking moments about and so doesn't contribute a moment. Each moment is the distance to the point where the force acts multiplied by the component of the force perpendicular to the line joining that point to the feet. These are shown in the figure below.

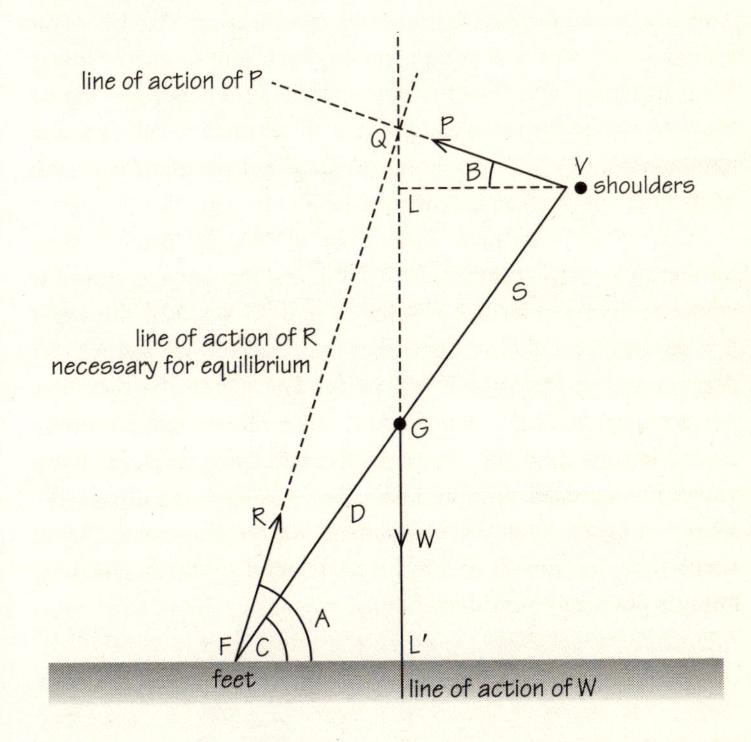

We shall label the distance from the surfer's feet to his centre of mass by D and the distance, GV, from his centre of mass to his armpit by S. The condition that the moment of his weight exactly counterbalances the moment of the arm pull in the opposite direction is simply that:[1]

$$Mg \times D\cos C = P \times (D+S)\sin(B+C)$$

If we combine this with our formula for P in terms of the weight then:[2]

$$\tan A = (S/D)\tan B + (1+S/D)\tan C$$

This is a useful formula because we can measure S and D for a typical windsurfer and gauge the angles B and C from filming them in action. This enables us to calculate the angle A – via its tangent – using the formula, which is all we need to calculate the denominator in our first formula (*) for P that we found from the horizontal and vertical balances of forces.

If our typical windsurfer has a mass of 70kg, the distance from his feet to his mass centre is D = 1.5m, and the distance from his centre to his shoulder is S = 0.5m, then S/D = 1/3. If the angle B = 45 degrees and C = 30 degrees then[3] tanA = 1.5 and A = 56 degrees and so the force P = 0.56Mg. The windsurfer therefore needs to supply a force that is about 56% of his weight for a long period of time (probably significantly more when wind and wave variations are taken into account). For a 70kg surfer this is the weight of about 40kg – for comparison, the weight limit for a free checked bag on British Airways is 23kg – and so this fun-looking event is physically very demanding.*

*Further details of the aerodynamics of the sail can be found in M. S. Townend, *Mathematics in Sport*, Ellis Horwood, Chichester (1984), pp126–7.

53

Winning Medals

Sport has long been a weapon in the battle for international prestige and one-up(wo)manship. Although the excesses of state-controlled sport in the East European communist states is a thing of the past, there is still a massive desire amongst wealthy nations to be top of the medals table – especially if you are the host nation and can field competitors in every event more cheaply than countries transporting their team members from the other side of the world. Suppose you were appointed head coach, handed a big cheque, and told to win as many medals as possible at the Olympic Games for your country in four or eight years' time. What should you do? If you look at the way in which China prepared for the Beijing Olympics, with what seemed to be an unlimited budget, you get some clues.

First, you might want to target sports where comparatively few countries compete – diving, rowing and cycling are better bets than football, boxing or athletics. Often, the unpopular sports owe their position to the expense of providing facilities (for example, high-board diving platforms and rowing eights) and so you will be able to press home your advantage with the money you have. Others, however, are unpopular for cultural or climatic reasons. There are few African swimmers, Nepalese sailors, Jamaican skiers or female runners and swimmers from Islamic countries.

Next, and most important, you need to start thinking about what it takes to succeed in different events. In all cases, you need a mix of natural talent and hard work. All you can do on the natural talent side is set up a scheme that will allow it to be identified and nurtured. But if you need to run 9.5s to win the 100m or 3m 26s to win the 1,500m your scouting system will be very unlikely to be successful: the level of natural talent required to compete at the same level as Usain Bolt or Hicham El Guerrouj is too high. So your strategy for winning medals should be quite different. Pick on the events where hard work, well-planned training and teamwork really pay off. In Beijing we saw the great successes by Chinese rowers and weightlifters. These are typical events where great strength can be enhanced and turned into medal-winning success. An added factor is that there are many medals available for events that are fairly similar in the basic strengths and skills required. Win the 100m freestyle swim and you will likely win the 50m or the 200m (and contribute to several relay teams) as well. In the case of rowing or cycling there is the possibility to create a very large squad covering many events where all the coaching expertise available benefits them all. Lots of training time is needed – possibly more than to win the 100m sprint – but you have lots of candidates.

Something similar has happened with Britain's success in track cycling. Again, this is a sport that rather few countries participate in. Indoor velodrome facilities are very expensive. But investment in coaching and better aerodynamic bikes and kit is cost effective: it benefits all the riders whatever their event. The need for special facilities means they have to train together and this helps engender team spirit and shared expertise. Athletics, by comparison, is a collection of completely different events: the skills needed to hurdle don't much help you throw the javelin or run a marathon. So, in your quest for lots of medals, you will be most interested in sports where there are many very similar events. Small details will change, like the distance to be negotiated or aspects of style or weight classes of competitors, but your coaching investment will help everyone. Weightlifting, boxing, judo and wrestling are all examples of this type of shared-expertise sport; athletics is not.

African runners provide us with another interesting example. They have a wide pool of natural talent helped by athletically active childhoods – no staring at computer screens all day – and (in some cases) a lifetime of living and running at high altitude. More important, they require no special equipment, no expensive facilities, just a good number of calories for their energy needs – and some other athletes to compete with. Last, but very important, is the absence of competition from other high-profile sports. This means that the best African athletes all try distance running and track running. In the UK we try to compete at a high level in a huge number of sports which results in a big dilution of talent. Why do we not have lots of world-class shot-putters and hammer throwers? All those people became rugby players or boxers; in the US they became American footballers. And the high jumpers were lost to basketball.

54

Why Are There Never World Records in Women's Athletics?

One of the key factors in attracting public attention, TV coverage, advertising and spectator numbers to sports is the occurrence of world records. This is why meeting promoters often put up large cash bonuses for any competitor who sets a world record (they take out insurance too). Some venues, like the Bislett Stadium in Norway, have an extraordinary history of world-record setting. Sometimes there can be aspects of a stadium that are advantageous – it might be at high altitude, for example, have a great crowd atmosphere or have favourable wind conditions.

It has long been appreciated by aficionados, but not shouted from the rooftops to the public at large, that there is a big difference between the chances of record-breaking in men's and women's athletic events. Here is a table showing the dates of the last world records in all the Olympic track and field events.

Event	Date of last men's record	Date of last women's record
100m	2009	1988
200m	2009	1988
400m	1999	1985
800m	2010	1983

1,500m	1998	1993
5,000m	2004	2008
10,000m	2005	1993
Marathon	2008	2003
110m hurdles	2008	1988
400m hurdles	1992	2003
3,000m steeplechase	2004	2008
4x100m relay	2011	1985
4x400m relay	1993	1988
Pole vault	1994	2009
High jump	1993	1987
Long jump	1991	1988
Triple jump	1995	1995
Javelin	1996	2008
Discus	1986	1988
Hammer	1986	2010
Shot	1990	1987
Decathlon/heptathlon	2001	1988

We see that the number of men's and women's records set at different times in the past is:

	< 5 yrs old	5–10 yrs old	10–15 yrs old	15–20 yrs old	> 20 yrs old
No. of men's records	7	3	4	6	2
No. of women's records	5	2	1	3	11

Looking behind the data we need to know that the women's 5,000m only entered the Olympic programme in 1995 and that the steeplechase, hammer and pole vault are also recent additions to the women's schedule of events. If we focus on the high-profile track events, like the sprints, hurdles and jumps that have always been there, then a significant fraction of the spectators (and competitors) at an athletics event might not have been alive when a world record was last set in those women's events!

If we look at the history of the long-jump records then the situation looks odder still. The record was last broken in 1988 but prior to that it was beaten six times between 1982–8 and four times in the two years before that. It seems clear that the key factor in the mystery of the long-standing women's world records is linked to the introduction of far stricter drug testing methods, and bans for offenders, in 1989. After the fall of communist East Germany, there were a series of investigations of Stasi secret police files by the German parliament in 1990 which uncovered detailed documentation of the systematic dosing of athletes with drugs in training and immediately before competition. Some athletes have claimed that they had no knowledge of what they were being given by coaches, and East Germany's most famous athlete, the sprinter and long jumper Heiki Drechsler, fought an unsuccessful libel action against the scientists who documented the fact that she had been given anabolic steroids throughout her career.[1] Consequently, some have argued that women's athletics might best be rejuvenated by ignoring records set more than twenty years ago and by simply beginning again. Only then, they argue, will the competitors of today and tomorrow be competing on a level playing field.

55

The Zigzag Run

In many team sports, like hockey, basketball, handball, soccer or rugby, there is scope for spectacular individual play. A fast-moving star player goes one way, then another, then another, taking the ball goalwards past a whole line of defenders. Stepping unpredictably to the right or to the left at each challenge produces a zigzag route towards goal. It is slower than the straight-line route that would be taken if there were no defenders in the way, but it offers the possibility of passing or completely avoiding defenders. Our attacker has three basic choices each time he encounters a defender: he can go straight on – perhaps a feint will make the defender go to the left or right or the attacker will just use his speed to run past the defender – or go approximately 45 degrees to the left or to the right.

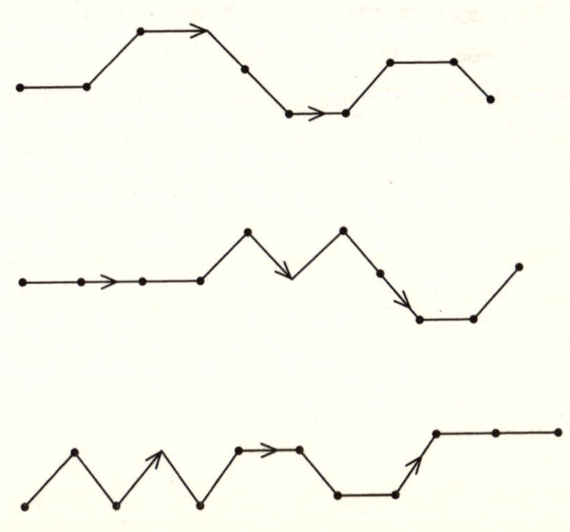

If our player was being entirely random about his or her mazy run then one of the three choices of direction is being made with a probability of 1/3 each time a defender is encountered. If D is the number of defenders encountered en route to the goal, or the try line, then the attacker has 3^D different zigzag routes to choose from. In soccer you might expect to encounter D = 8 defenders (including the goalkeeper) and you have a dizzying $3^8 = 531,441$ runs available to you. A rugby union player might have as many as $3^{14} = 10,460,353,200$ options available. When the number of encounters gets large the player's route through the opposing defence begins to look like a classic problem in physics called the random walk. In a random walk you choose the direction of your next step at random. A general feature of such a random walk is that you need N^2 strides to reach a distance equal to N stride lengths from your starting point. If each stride length is 1.5m and you are making a run over 45m at 9m/s then running in a straight line will take 5s and you will use 30 strides. If you make eight swerves it will take $30 + 8 = 38$ strides to get to the goal and this will take a time of $38 \times 1.5 \div 9 = 6.3$s. Tackling opponents have an extra 1.3s to catch you. If they run at 9m/s as well then they will cover 11.7m in that time.

Your random swerving can make you drift away from the target – the goal. Suppose a footballer started at the centre circle so that the goal is straight in front of him, 45m away. As he keeps on swerving right or left or not at all, he encounters nine defenders. If he responds at random, then he is likely to drift about $\sqrt{9} = 3$ stride lengths, or 4.5m, away from the centre line to goal.[1] The more tackles he has to evade (some of the defenders might return for a second challenge) the farther he is likely to drift as a result of his random swerves.

56

Cinderella Sports

There is a solid core of nine Olympic sports that have been there since the first Games in 1896. Others, like BMX biking and triathlon, have been added, while polo, croquet and lacrosse have gone, probably never to return. The Summer Games in 2012 will include twenty-six different sports with subdivisions into thirty-six separate disciplines and a total of 300 different events. Cycling, for instance, is divided into four disciplines: BMX, mountain biking, road cycling and track cycling. Aquatics is split into diving, water polo, synchronised swimming and swimming.

The International Olympic Committee (IOC) officially recognises a large number of other sports. Some of these 'Cinderellas' could be added to future Olympics if they garner the requisite number of votes from the IOC membership when they decide on changes to the Olympic programme. Yet, to be recognised by the IOC, they must follow the Olympic Charter's principles and have an overseeing federation that guarantees that they will continue to do so. Some of the recognised Cinderella sports, like chess and bridge, will never be admitted to the Olympics proper because they do not possess a physical dimension. Instead, they can be incorporated into parallel competitions like the World Games, which are held quadrennially under the patronage of the IOC and first took place in 1981. Some of the more physical Cinderella sports, like rugby sevens, are scheduled to enter the Games proper in Rio in 2016.[1]

At present, the list of IOC-recognised sports consists of the following:

Air sports	Bandy	Baseball	Billiard sports
Boules	Bowling	Bridge	Chess
Climbing	Cricket	Dance sports	Floorball
Golf	Karate	Korfball	Lifesaving
Motorcycling	Netball	Orienteering	Pelota vasca
Polo	Powerboating	Racquets	Roller sports
Rugby	Softball	Squash	Sub aqua
Sumo	Surfing	Tug of war	Water skiing
Wushu			

This is an interesting list and most readers will not have heard of all these sports. For those unfamiliar with it, bandy is a Russian version of outdoor ice hockey – a winter sport played with eleven a side on a frozen surface the size of a football pitch. The rules are patterned to some extent on football. Korfball is a hybrid of netball and basketball that is very popular in Holland and Belgium, and floorball is just indoor hockey. Pelota vasca is an ancient Basque precursor of tennis, played against a wall, like squash, but on a very large court (more than 50m long) where the players use a slinglike scoop to catch and throw a leather ball with enormous velocities (sometimes exceeding 180mph). Wushu is a Chinese martial art that dates from 1949. It can either be performed solo with points scored for style and technique – like figure skating – or in bouts between two competitors.

I can't help thinking that tug of war would become the most popular team event at the Olympics if it was readmitted. It was a team sport in the Olympic Games from 1900 to 1920 and Britain, the last gold medallists, tops the medal table with two golds, two silvers and a bronze. However, only three nations entered teams between 1900 and 1908; and this fell to two in 1912, before rallying to five in 1920. So its exit from the Games was not surprising. But maybe a tug in the right direction could bring it back.

57

Wheelchair Racing

The most spectacular Paralympic races are the wheelchair events on the track. Athletes are graded into competition classes according to the nature of their disability. Competition is ferocious, especially in events like the 800m which are not raced in lanes. Arguably, the most successful individual Olympic competitor in history is the 39-year-old Canadian multiple medallist, Chantal Petitclerc, who won five gold medals and set three world records in wheelchair racing at the Beijing Paralympics, giving her an amazing total of fourteen gold, five silver and two bronze individual event medals since 1992. Britain's Tanni Grey-Thompson is not far behind with eleven gold, three silver and one bronze individual event medals from five consecutive Olympics.

As one might expect, there are a number of detailed stipulations about competition wheelchairs. Some are for reasons of safety – no mirrors, and no front, back or side protrusions; others are to ensure fairness – no mechanical steering devices, no gears or levers. There are limits on the maximum height that the main body of the chair can be above ground (50cm). This is for safety – too high and it can overturn very easily. The larger rear wheel can't exceed 70cm and the front wheel can't exceed 50cm in diameter when the tyres are fully inflated. The wheel diameter determines the speed at which the wheels will turn when the hand rim is spun around the wheel and so it has been standardised.

These rules about the wheelchair structure have an unexpected omission. There is no stipulation about its weight.[1] Competition wheelchairs can vary in mass between 6–10kg but the best new models, being introduced by Franz Fuss and his engineering team in Melbourne, weigh as little as 5kg.[2] If you add the weight of the competitor then the variation could be as great as 20kg within one event.[3]

Why is wheelchair mass so important? Wheelchair racing is like cycle racing in which motion is resisted by two forces: air resistance and rolling friction of the wheels with the ground. Wheelchairs go more slowly than bicycles (no faster than 10m/s, which is the finishing speed in the world record 100m wheelchair event) and they weigh a lot more. Interestingly, the mass of racing bicycles is limited by International Cycling Union regulations to be no less than 6.8kg.

The different speeds and weights mean that the relative importance of air drag and rolling friction is different for wheelchair motion. The drag force on the moving wheelchair and athlete is:

$$F_{drag} = \tfrac{1}{2} CA\rho v^2$$

where $\rho = 1.3\text{kg/m}^3$ is the density of air, A is the cross-sectional area that the wheelchair and the athlete presents to the air, C is the so-called drag coefficient that reflects how aerodynamic and smooth the moving body is, and v is the speed of motion of the wheelchair relative to the air. Notice that this drag force doesn't depend directly on the mass of the chair and the athlete. Some of these factors are under the athlete's control. You can shift your body position so that the area presented to the air is reduced and you can wear kit that reduces C by avoiding rough fabrics or flapping sleeves. The effective area that causes resistance is the combination CA and for wheelchair athletes this tends to be about 0.14m^2. So, the drag force at a speed of 10m/s is about $F_{drag} = 0.07 \times 1.3 \times 100 = 9.1\text{N}$.

The other force slowing the wheelchair down is the frictional resistance to the rolling wheels. This is proportional to the total weight of the chair plus its rider, Mg, acting vertically downwards:

$$F_{roll} = \mu Mg$$

where μ is the coefficient of friction which is typically about 0.01 for a wheelchair. Unlike the drag force, it is independent of the speed.[4] For a wheelchair carrying an athlete of mass 80kg, and acceleration due to gravity of $9.8m/s^2$, this frictional force is typically $F_{roll} = 0.01 \times 80 \times 9.8 = 7.8N$. This is very similar to the air-drag force but for most of the race the wheelchair will be travelling much slower than its top speed of 10m/s and the rolling friction will be the largest resistance for the athletes to overcome. For example, over 100m, during the world record time of 13.76s, the average speed is 7.3m/s and the air-drag force at this speed is smaller by $(7.3/10)^2$, and drops to 4.8N, much less than the rolling frictional force.

This shows that the rolling friction is very important and it is determined by the weight of the chair plus the athlete. Athletes who are able to use high-tech wheelchairs made out of strong lightweight materials are at a significant advantage. Franz Fuss's laboratory in Melbourne has done detailed experimental studies of the effects of the chair weight on performance.[5] They confirm that saving 1kg or 5kg from the mass of the wheelchair would improve 100m times by about 0.1s and 0.6s respectively, for a representative 60kg male athlete. The effects on longer races will be correspondingly more significant. We would expect that advancing technology will soon lead all top wheelchair racers to have the lightest available wheelchairs and the variance in performance created by wheelchair mass differences will be removed.

What seems to have been missed in these studies is the fact that variations in the weights of the athletes play a much more significant role than the weights of the wheelchairs and are much more variable from competitor to competitor. Clearly, it pays to have a lower body weight so as to reduce the rolling friction. It is a very costly and time-consuming project to reduce the weight of your racing wheelchair by 3kg, but much cheaper and easier to reduce your weight by the same amount. It might even be argued that there should be weight categories for wheelchair competition, but this would lead to an impractical proliferation of events and weight classes. A far easier method would be to determine the weight of each competitor plus their wheelchair before the start and then add weights so that each one has a predetermined standard weight. I wonder if it will ever happen.

58

The Equitempered Triathlon

The triathlon is the newest sport in the Olympic Games. Despite appearing first in the 2000 Games in Sydney, it was only invented in its modern form in 1978 by a group of runners from the San Diego Track Club who staged an event for forty-six intrepid souls who wanted to subject themselves to a swim, a bike ride and a distance run with no rest between.[1] Later that year, an even fiercer 'Ironman' test was created in Honolulu, with a 2.4-mile (3.9km) swim plus a 112-mile (180.2km) ride and a marathon (42.2km) run to finish. Remarkably, all but three of the fifteen starters finished, led home by Gordon Haller just 11h 46m and 58s after he started. There are several versions of the triathlon now (the shortest being the 'Sprint' event with a 750m swim, 20km ride and 5km run) but we will confine our attention to the standard Olympic event. For both men and women, it starts with a 1.5km swim, then a 40km bike ride, followed by a 10km road run. The winner is determined by the first to the overall finish line, that is by the shortest time when the times for the swim, ride and run are added together with the two (short) transition times needed to switch from one discipline to the next.

The times for each of the Beijing medal winners is shown in the two tables below, with the split times for each of the three disciplines, together with the fastest time recorded for each stage by any competitor (note the first three times don't add to the total because the two transition times must be added).

The obvious question to ask about this event is whether the

relative lengths of the swim, ride and run stages are fairly chosen. Some triathletes are primarily runners, others naturally swimmers or cyclists, and the amount of time they have to exploit any advantage they have on their strongest discipline at the expense of their weakest is crucial for their overall performance. As the rules stand, the winning man spent a mere 16.7% of his total time swimming, 28.3% of it running, 0.8% of it in transitions, and a whopping 54.2% of it cycling. The fractions of time spent on the different legs are roughly the same as this for the women and the other men.

These numbers appear rather shocking. There is clearly an imbalance in this event, with far too much weight being placed upon the performance in the cycling. An outstanding cyclist has more time to exploit his strength than the runner and swimmers combined.

Men	Swim	Ride	Run	Overall time
Jan Frodeno	18m 14s	59m 1s	30m 46s	1hr 48m 53s
Simon Whitfield	18m 18s	58m 56s	30m 48s	1hr 48m 47s
Beven Docherty	18m 23s	58m 51s	30m 57s	1hr 49m 5s
Fastest overall	18m 2s	57m 48s	30m 46s	

Women	Swim	Ride	Run	Overall time
Emma Snowsill	19m 51s	1hr 4m 20s	33m 17s	1hr 58m 27s
Vanessa Fernandes	19m 53s	1hr 4m 18s	34m 21s	1hr 59m 34s
Emma Moffatt	19m 55s	1hr 4m 12s	34m 46s	1hr 59m 55s
Fastest overall	19m 49s	1hr 3m 54s	33m 17s	

In order to make the event fairer and more attractive to strong swimmers and runners, the stage lengths could be equalised – but that would not be sensible: cycling is faster than running and running is faster than swimming. Instead, it would be best to equalise the time spent on each discipline.[2] This is fairest because it is the total overall time that determines the winner. This could be done by making the swim and the bike legs of the same duration as the run. Better still would be to keep the total time roughly the same – say 1hr 48m – and have equal time on each of the three stages. This would give thirty-six minutes for each stage and the distances could be 3km for the swim, 24km for the bike ride and 12km for the run. Alas, most triathletes would be aghast at the extra length of the swim. Swimming is notoriously difficult because athletes need to work on technique and swimming efficiency. It is not enough just to do more training like you do for the bike ride or the run; if your swimming technique is poor then training will just end up ingraining it deeper and make it harder to repair.

Yet this would be the perfectly tempered triathlon. I recommend it for future Olympic Games. It is a fairer all-round test than the event we have.

59

The Madness of Crowds

If you have ever been in a huge crowd, at a sports event, a pop concert or a demonstration, then you may have experienced or witnessed some of the strange features of collective behaviour. The crowd is not being organised as a whole. Everyone responds to what is going on next to them, but nonetheless the crowd can suddenly change its behaviour over a large area – sometimes with disastrous results. A gentle plodding procession can turn into a panic-stricken crush with people trying to move in all directions. Understanding these dynamics is important. If a fire breaks out or an explosion occurs near a large crowd, how will they behave? What sort of escape routes and general exits should be planned in large stadiums? How should religious pilgrimages of millions of worshippers to Mecca be organised so as to avoid repeating the deaths of hundreds of pilgrims that have occurred in the past, as panic generates a human stampede in response to overcrowding?

One of the interesting insights that inform studies of crowd behaviour and control is the analogy between the flow of a crowd and the flow of a liquid. At first one might think that understanding crowds of different people, all with different potential responses to a situation, different ages and degrees of understanding of the situation, would be a hopeless task. Surprisingly, this is not the case. People are more alike than we might imagine. Simple local choices can quickly result in an overall order in a crowded situation. When you arrive at one of London's big rail termini and head down to the Underground system, you will find that people descending will have chosen the left- (or right-) hand stair, while those ascending will keep to the other side. Along the corridors to the ticket barriers the crowd will organise itself into two separate streams moving in opposite directions. Individuals take their cue from what they observe in their close vicinity. This means that they act in response to how people move nearby and how crowded it gets. Responses to the second factor depend a lot on who you are. If you are a Japanese manager used to travelling through the rush hour on the Tokyo train system you will respond very differently to a crush of people around you than if you are a tourist from the Scottish Isles, or a school group from China. If you are looking after very young or old relatives then you will move in a different way, linked to them and watching where they are. All these variables can be taught to computers that are then able to simulate what will happen when crowds congregate in different sorts of space, and how they react to the development of new pressures.

Crowds seem to have three phases of behaviour, just like a flowing liquid. When the crowding is not very great and the movement of the crowd is steady in one direction – like the crowd leaving Wembley Stadium for Wembley Park Underground station after a football match – it behaves like a smooth flow of a liquid. The crowd keeps moving at about the same speed all the time and there is no stopping and starting. As the density of people in the crowd grows significantly, they start pushing against one another and movement starts to occur in different directions. The overall movement becomes more staccato in character, with stop-go behaviour, rather like a succession of rolling waves. The gradual increase in the density of bodies will reduce the speed at which they can move forward and there will be attempts to move sideways by those who sense that things might move forwards faster that way. It is exactly the same psychology as cars swopping lanes in a dense slow-moving traffic jam. In both cases it sends ripples through the jam which causes some people to slow and others to shift sideways and let you in. A succession of those staccato waves will run through the crowd. They are not in themselves necessarily dangerous but they signal the possibility that something very dangerous could suddenly happen. The closer and closer packing of people in a crowd starts to make them behave in a much more chaotic fashion, like a flowing liquid becoming turbulent, as people try to move in any direction so as to find space. They push their neighbours and become more vigorous in their attempts to create some personal space. This increases the risk of people falling, becoming crushed together so closely that breathing is difficult, or children getting detached from their parents. These effects can start in different places in a big crowd and their effects will spread quickly. The situation rapidly snowballs out of control. The fallers become obstacles over which others fall. Anyone with claustrophobia will panic very quickly and react even more violently to close neighbours. Unless some type of organised intervention occurs to separate different parts of the crowd and reduce the density of people, a disaster is now imminent.

The transition from smooth pedestrian flow to staccato movement and then crowd chaos can take anything from a few minutes to half an hour depending on the size of the crowd. It is not possible to predict if and when a crisis is going to occur in a particular crowd. But by monitoring the large-scale behaviour of a crowd, the transition to the staccato movement can be spotted in different parts of it and steps taken to alleviate crowding at the key pressure points that are driving the transition where chaos will set in.

60

Hydrophobic Polyurethane Swimsuits

The world of swimming has recently emerged out of a very difficult period in which it had to come to grips with the role of new technology in the sport. We are used to technical improvements in pieces of equipment like tennis rackets, fibreglass vaulters' poles or golf clubs changing performance levels, but the appearance of whole-body polyurethane swimsuits has taken the issue to another level. Swimmers have always taken steps to reduce the drag on their bodies through the water. All body hair was shaved off ahead of major competitions and sleek swimming caps were used to remove any drag from head hair. But the new swimsuits took these measures a whole lot further. They were made from an extremely thin layer of foam-like material that enclosed tiny pockets of gas that made the swimmer wearing the suit far more buoyant. As a result swimmers floated higher in the water and were subject to less drag. The suits in effect pushed water away from the swimmer's body and were therefore dubbed 'hydrophobic'.

The drag on the human body moving in water is around 780 times larger than the drag when moving in air and so there is considerable advantage to getting as much of the body above the water level as possible. These suits also made the body shape very smooth and hydrodynamic. Instead of the joint between a man's body and the waist cord of his swimsuit adding extra drag, there was now a seamless, wrinkle-free, low-resistance outer shell skimming through the water. Tiny fibres on the surface of the suit could move to keep the shape streamlined and its texture smooth as the body shape changed through the stroke. Overall there was the possibility of an 8% reduction of drag on a swimmer. There are downsides, however. Putting on one of these thin-film polyurethane suits takes about half an hour, so you wouldn't want to use one in every early morning training session! And they don't last long: you will need a new one after every few races and they are not cheap, costing about $500 each in America.

The result of all of this was the erasure of outstanding world records by performances that were intrinsically inferior. Twenty new world records were set at the World Swimming Championships in Rome during July 2009 alone. Not all swimmers were wearing these suits at championships, though, and races were becoming manifestly unfair. Those who wore them were entering a technological arms race as different sponsoring companies tried to produce a superior suit for their swimmers. Moreover, the sponsorship deals that top swimmers had entered into prevented them from switching to the best suit if it was made by a rival company.

The world's best swimmer, Michael Phelps, seemed to concentrate everybody's minds when he suggested, via his coach Bob Bowman, that he might boycott all future international competitions where the new suits were allowed because they were distorting the sport. Swimming and the IOC seemed to be faced with a future without a competitor who had won fourteen Olympic gold medals – more than any other sportsman or woman in history.

Not surprisingly, then, in 2010 there was a ban on buoyant polyurethane swimsuits. An uncompromising American proposal to return to allowing only textile suits was passed with 180 nations voting for it and only seven against. While FINA didn't nullify past records that were set by athletes using polyurethane suits, they were 'starred' in the same way that records set at high altitude are marked separately in the record books of track and field.

61

Modern Pentathlon

The modern pentathlon is a diverse challenge of all-round sporting ability. It was invented by the founder of the modern Olympics, Baron Pierre de Coubertin, but it didn't enter the Games as a men's event until 1912; the women's event didn't appear until 2000, when it was won by Stephanie Cook. The 1912 event featured six Swedes amongst the first seven finishers. The odd one out was the American army officer George Patton, who was to become the famous commander of the US Third Army during World War II. Patton finished fifth but argued that he should have been the gold medallist. He was judged to have missed the target with one of his shots but he claimed that the bullet had passed through the hole of an earlier shot and so hadn't left a mark. The judges ruled against him.

The men's champion at the last two Olympics has been the Russian Andrei Moiseev, who finished far ahead of his rivals in Athens and Beijing. Some of the five sports involved – shooting, fencing, riding, swimming and running – are divorced from the experience of most young people, and were chosen originally to represent the skills that an ideal cavalry soldier of the late nineteenth century would need to survive behind enemy lines. He might need to ride a strange horse, fence or shoot his way out of trouble, or simply run or swim for it if all else failed!

The Olympic event has an interesting specification. Scoring is based on a reference guide score of 1,000 points per event. A predetermined standard is given for each event and the point score you receive is determined by how much better (or worse) you do than that standard. The fencing competition, with épées, is a round robin – each competitor must face every other – and one touch wins the match. The reference standard of 1,000 points is gained by winning 70% of your matches. Each win or loss above or below is worth a number of points depending on the number of competitors (for example, ± 40 points when you contest twenty-two or twenty-three matches). Shooting is now with a 4.5mm air pistol (which will be replaced by a laser device in 2012), with twenty shots from a ten-metre distance and a maximum of 10 points for each shot, giving a total score of 200. You gain 12 points for each point you score above a standard shooting total of 172. The swim is 200m freestyle with a standard of 2m 30s for men and 2m 40s for women. Each third of a second above or below this standard score results in 4 points deducted or added from the standard 1,000. The riding is the real wild card. The horses are allocated to riders at random from a group and you have twenty minutes to familiarise yourself with your horse. Then you have to negotiate twelve to fifteen jumps on a show-jumping course approximately 400m long. A fault-free ride inside the time limit gives 1,200 points but you lose 28 points for knocking down a fence, 40 for a refusal, 60 for disobedience that knocks down an obstacle, and 80 for a missed or refused obstacle. The 3km cross-country, or road, run is the final event and the points already accumulated determine when you start. The standard time is ten minutes for men and 11m 20s for women with each one-second difference counting 4 points. The competition leader starts first and the intervals between runners are staggered to reflect the overall points differences after the previous four events – each point difference gives a one-second start time difference – so that the finishing order at the end of the run will be the overall finishing order for the whole pentathlon.

A radical rule change has recently taken place which will affect the 2012 competitions. Shooting and running have been combined so that the pentathletes will do their shooting at intervals during the run: at the start, and then after 1,000m and 2,000m of the 3,000m run. Each time they will shoot at five targets and have only seventy seconds for their shots or incur penalties; the best shooters will take about thirty seconds for their five shots. This dramatically changes the nature of the competition and it becomes more like the skiing and shooting biathlon at the Winter Olympics, although the runners will not carry their guns like the skiers do. Shooting requires a calm and steady hand with a reduced heart rate. Shooting after you have been running is therefore tricky and there is a trade-off between running hard and being able to shoot well at the 1,000m and 2,000m intervals. The overall winner will still be the first person that crosses the 3,000m finish. Of course, this is no longer quite a pentathlon. It is perhaps a triathlon plus a biathlon.

The top pentathletes are very good swimmers and the best male and female swimmers will go faster than 1m 55s and 2m 8s respectively for their 200m swim. It is clear that a small amount of tinkering with the points scheme for the swim would be more even-handed. Swimming standards have advanced dramatically over the years in the specialist swimming events and this has carried over into the pentathlon – Andrei Moiseev is a very strong swimmer and scored 1,376 points in his Beijing swim but no more than 1,036 points in any of the other four disciplines. Fencing is significantly undervalued and was the only sport where the average score of both the top six men and women in Beijing fell below the 1,000-point standard (888 for the women and 920 for the men). Ironically, the new rule changes which merge the shooting and running leave these two anomalies in the scoring system unchanged.

	Shoot	Fence	Swim	Ride	Run
Men: % of total points score	21	16	24	20	19
Women: % of total points score	20	16	22	20	22

62

Keeping Cool

There has been a growing tendency for major sports championships to be held in hot countries during the summer. Athletes are generally not consulted about these decisions and so find themselves playing football in Qatar, running marathons in Athens or riding horses in Seoul. At least this won't be a problem in London in 2012.

The longest battles against the heat are in the road-walking races over 50km and in the marathon run. Considerable efforts are taken by the competitors to remain hydrated throughout these races and there are regular drinks stations around the courses to take on water or electrolyte drinks. It is important to start taking these at the earliest opportunity: if you wait until you start feeling thirsty then it's too late. We have also seen competitors experiment with different types of clothing to keep cooler before and during the race; before the early morning start to the women's marathon at the Athens Olympics, Paula Radcliffe was wearing a strap-on rubber jacket with pockets filled with ice. Some runners wear string vests that allow a maximum airflow through to the body – although they mean you also feel more sunlight – while others have tried reflective materials that minimise the absorption of the sun's energy. All serious competitors who are not based in hot climates will have invested a lot of effort training in hot conditions so as to acclimatise their bodies to extreme heat and accurately gauge how much liquid they need to drink before and during the race. The most famous case of this sort was the remarkable British race walker, Don Thompson, who won the gold medal in the 50km walk in very hot conditions at the 1960 Rome Olympics. He prepared for the heat by exercising in his bathroom with the temperature raised to over 100°F (38°C) using wall heaters and kettles of boiling water to raise the humidity. He even had a stove in his bath! In those days there was no athletic sponsorship and Thompson worked nine to five as a fire insurance clerk for Commercial Union. He had to get up each morning at four in order to get in his long training sessions before changing for work. But at the Rome Olympics his preparations really paid off. Temperatures soared above 30°C during the 50km walk and his rivals wilted one by one. He took the lead at halfway in his sunglasses and Foreign Legion cap and was never headed, winning by seventeen seconds in an Olympic record time. The Italian press dubbed him 'Il Topolino' – the Mighty Mouse. Usually, he wrote only one line each day in his carefully kept diary: that day he wrote two.

Are there any physical characteristics, like body size, that might help or hinder you in very hot conditions like those that Thompson encountered? Suppose that you are moving along at a steady speed V in a distance running or walking event. Then the heat you generate will be roughly equal to the energy of motion, $\frac{1}{2}mV^2$, needed to propel you forwards, where m is your mass. In order to stay in thermal balance and not overheat, your body will need to be able to cool at the same rate. Cooling is proportional to the surface area of your body, A, so when you are running or walking in equilibrium you should have cooling equal to heating and so $mV^2 \propto A$. Your mass is just your density times your volume. Your body density is fixed so the maximum steady walking speed is determined by the ratio of your surface area to volume ratio via $V^2 \propto$ area/volume. Suppose we model the body as a cylinder of radius R and height h whose bottom half is divided into two 'legs' of radius R/2 and height h/2. The total volume will be $\frac{3}{4}hR^2$.[1] The total surface area of the top and the outside surface (excluding the bottom surfaces of the two cylinders, 'the feet', because they will be pushing against the friction of the ground) will be $\pi R(R + 2h)$.[2] So, equating the heating to the cooling we have for the square of your comfortable speed:

$$V^2 \propto 1/h + 2/R$$

The message is that if you are shorter (smaller h) and thinner (smaller R) then you will find it easier to keep cool when racing in hot conditions. We can also see that h is much bigger than R in practice so V is determined more by your body radius, R (your waist measurement is $2\pi R$), than by your height. Small athletes do have an advantage in hot conditions.

63

Wheelchair Speeds

The first international sports events for disabled athletes held in conjunction with an Olympic Games were in Rome in the summer of 1960. The term 'Paralympics' was only coined four years later at the Tokyo Olympics. At first, wheelchair athletes used conventional heavy wheelchairs (7–18kg) and only contested events up to 200m. The first wheelchair-borne marathon competitor competed at the 1975 Games in Boston. New races were organised and gradually purpose-built racing chairs appeared. By the 1980s they had become lightweight and technically sophisticated. The first sub-four-minute mile by a wheelchair athlete came in 1985 and intense competition has driven down records in events on the track and in the marathon.

It is interesting to look at the trends in world record performances for able-bodied and para-athletes. They are both very well defined but quite different. Able-bodied athletes are faster up to about 400m but after that their average speed quickly falls behind the wheelchair performances.

The two tables I have compiled show the world record times for the Olympic running and wheelchair events, for both men and women, together with the average speed of the athlete in each case – this is just the distance covered divided by the time recorded. If you like to think of speeds in miles per hour rather than metres per second, then 10m/s corresponds to 22.4mph, so Haile Gebrselassie cruises around the marathon course at about 12.7mph.

Men's event	Record time	Speed m/s	Women's event	Record time	Speed m/s
100m	9.58s	10.44	100m	10.49s	9.53
200m	19.19s	10.42	200m	21.34s	9.37
400m	43.18s	9.26	400m	47.6s	8.4
800m	1m 41.01s	7.92	800m	1m 53.28s	7.06
1,500m	3m 26s	7.28	1,500m	3m 50.46s	6.51
5,000m	12m 37.35s	6.60	5,000m	14m 11.15s	5.87
10,000m	26m 17.53s	6.34	10,000m	29m 31.78s	5.64
Marathon	2hr 3m 2s	5.72	Marathon	2hr 15m 25s	5.19

Men's wheelchair event	Record time	Speed m/s m/s	Women's wheelchair event	Record time	Speed m/s
100m	13.76s	7.27	100m	15.91s	6.29
200m	24.18s	8.27	200m	27.52s	7.27
400m	45.07s	8.88	400m	51.91s	7.71
800m	1m 32.17s	8.68	800m	1m 45.19s	7.61
1,500m	2m 55.72s	8.54	1,500m	3m 24.23s	7.34
5,000m	9m 54.82s	8.41	5,000m	11m 39.43s	7.15
10,000m	20m 25.9s	8.16	10,000m	24m 21.64s	6.84
Marathon	1hr 20m 14s	8.77	Marathon	1hr 38m 29s	7.14

You would expect to find that the average speed of each record run falls as the distance of the event gets longer. For able-bodied running records the trend is remarkably well defined and in the graph below you can see the systematic trends for the men's and the women's events. The average speed, U, in m/s varies with the distance, L, in metres as a power law with slope of about –0.1, and more precisely as $U \propto L^{-0.109}$ for men and $U \propto L^{-0.111}$ for women.

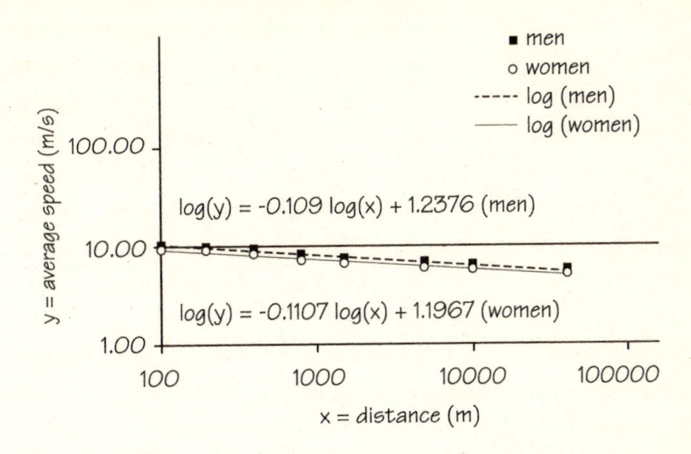

The corresponding results for the average speed trends in the wheelchair events are strikingly different. If we don't pay too much attention to the 100m event because of the strong effect that getting started and building up speed has on the total time, we can see already from the table of average speeds that there is hardly any fall-off of speed with distance. The athlete's technique reaches the fastest turnover rate quite quickly and can maintain it over very long distances. The proportionalities are very close to being flat, with $U \propto L^{-0.006}$ for men, $U \propto L^{-0.021}$ for women.

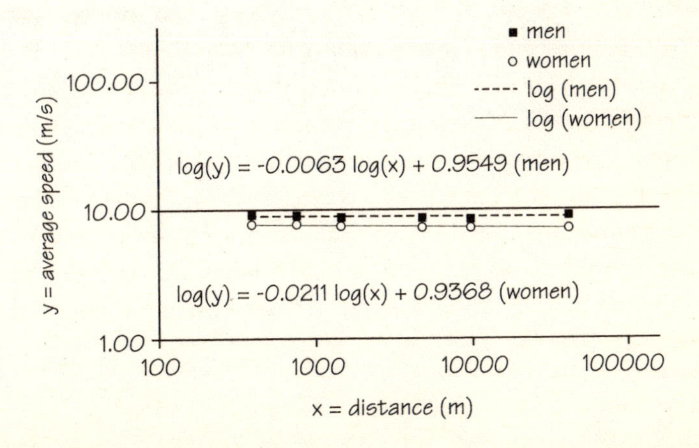

We can also see from the tables that the average speed achieved during the marathon record is *higher* than in the shorter distances like the 10km, and even the 5km. There are several reasons that contribute to this anomaly. The shorter-distance records are set on the track and require the negotiation of two bends on each 400m circuit of the track. Able-bodied runners are not much affected by this but it is more problematic for wheelchairs; it is also harder for them to overtake. Hence the wheelchair slows on the bends compared with the two straights. The marathon is raced over flat road courses with a minimum of twists and turns and is much better for wheelchair racing, although in practice the courses are optimised for the able-bodied runners. In addition, wheelchair marathons are more competitive and frequent than 10km track races and the records are under far more pressure with larger fields of competitors. The tactical nature of 10km track races also often leads to slower times.

The unusually small variation of speed with distance also allows wheelchair athletes to be competitive over a far greater range of distances than able-bodied athletes. David Weir has won the London marathon wheelchair race on five occasions but also has Olympic medals for the 100m, 200m, 400m, 800m, 1,500m and 5,000m wheelchair events. No able-bodied could ever hope to succeed in more than three of those events. Our average speed analysis shows why it is a possibility for para-athletes.

64

The War on Error

Errors cause problems. But errors are not necessarily mistakes. They are uncertainties in our knowledge of the true state of affairs. Sometimes these are one-offs: we want to measure a baby's weight and our digital scales are only accurate to a gram so we have an unavoidable imprecision or measurement 'error' in our determination of the baby's weight. Sometimes errors can be dramatically cumulative, and double (or worse) at each stage in a many-step process; this is the source of the phenomenon of 'chaos' that has been so well publicised over the past thirty years. It bedevils our attempts to predict the weather with great accuracy. In between these two extremes there is another type of error that stays the same at each stage in a many-step process but accumulates in ways that can lead to a significant overall uncertainty.

If you are building a swimming pool or a running track that is going to be used for races and time trials covering many laps, then the accuracy of the lap length is very important. Make it under-length and the race length will get more and more under-length as the length of the race increases. Any records set will be invalid when ultimately checked by laser ranging. Adjustments to the finish line of a running track can be made to counter final construction errors, but that option is not available in a swimming pool.

Suppose a race has length R and covers N laps of length L, so $NL = R$. If the construction of the venue has led to an error in the length of a lap equal to ε then the cumulative error over the race will be a distance $N\varepsilon = \varepsilon R/L$.

In practice, it is times that we are interested in for record purposes. If the time to complete the race is T then the average speed is L/T and the overall time error, ΔT, introduced by the error in the length of the lap is $R/L \times \varepsilon \div R/T = T\,\varepsilon/L$, so we have the very simple result $\Delta T/T = \varepsilon/L$. This is telling us that the fractional error in the total race time is equal to the fractional error in the lap length.

In the construction of international athletics tracks the IAAF tolerance standard is that a 400m lap determined by an average of measures of its straights and bends is allowed to be up to 4cm over-length but not under-length. For FINA-approved swimming pools, a 50m pool can be built up to 3cm over-length but there is no tolerance for it to be under-length. This is to ensure that records can't be broken by under-length races – or, worse still, that all times recorded on a short track or in a short pool become invalid for any statistical or comparative ranking purpose.

	Lap length. L	Max. allowed lap length error, ε	Overall race time error, ΔT, in secs if race lasts T secs	Duration of race with > 0.01s error
Athletics track	400m	0.04m	$10^{-4} \times (T/1s)$	100s
Long-course swimming	50m	0.03m	$6 \times 10^{-4} \times (T/1s)$	16.7s
Short-course swimming	25m	0.03m	$12 \times 10^{-4} \times (T/1s)$	8.3s

In the table we have worked out the consequences of this formula if the maximum tolerance is used and the running track is 4cm over-length and the swimming pool is 3cm over-length. In the last column, we work out the amount of time the race would have to go on for in order for this lap-length error to produce an overall timing error exceeding 0.01s, which is the accuracy of timekeeping for records in athletics and swimming. All swimming races exceeding 50m are affected as well as all standard track events of 800m and longer. For a 10,000m race at world record pace for men (26m 30s), the cumulative error would be 0.16s and therefore quite significant. Of course, the situation in running events is less clear-cut because competitors may run different total distances if they run wide of the inside-lane kerb for tactical reasons and in fact everyone runs over distance. In swimming the situation is more clear-cut. Careful surveying really counts.

Finally, there are human errors that no amount of planning can eradicate completely. In the 1932 Olympics an official lost count of the laps that had been run in the 3,000m steeplechase on the track and waved everyone round for an extra lap. The athlete in second place after 3,000m ended up in third place at the end of the extra lap. Oh dear.

65

Matters of Gravity

Many sports are constrained by the force of gravity. Competitors attempt to beat its effects when they throw objects further or jump higher and longer than their rivals. The range of a projectile or the height of a jump is inversely proportional to the acceleration due to gravity, $g = 9.8m/s^2$ on average on the earth's surface (g is six times smaller on the moon, incidentally).[1] The weight of a 100kg mass is given by the mass and g. But the value of g varies according to where you are located on the earth's surface. This is a result of two effects. The first, and less important, is that the earth is not perfectly spherical.[2] It is slightly fatter at the equator than at the poles and this means that the distance to the centre of the earth increases as you move from the poles to the equator – although not quite steadily because there are occasional mountains and gorges along the way. The more significant effect on the strength of the gravitational pull we feel is created by the daily rotation of the earth about its axis.

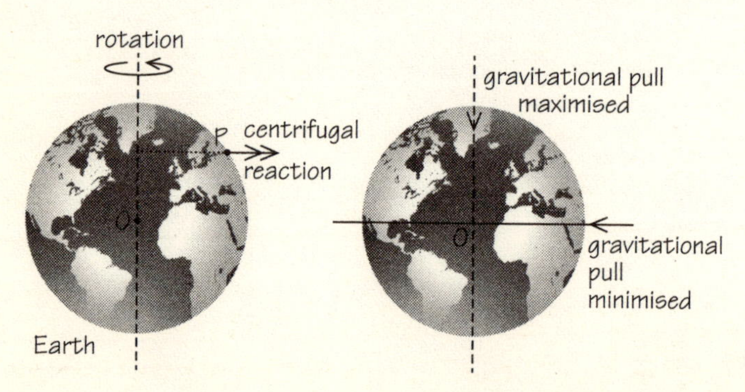

The effect of the rotation on someone standing at the earth's surface is to rotate them around in a circle that increases steadily in radius from zero at the poles up to its maximum at the equator. This creates an outward centrifugal force at right angles to the earth's rotation axis. This force opposes the pull of gravity created by the mass of the earth so that a spring balance with a 1kg mass hanging from it records a weight that is a maximum at the poles but falls steadily to a minimum as you move it to the equator:[3]

$$g(\text{North and South Poles}) > g(\text{equator}).$$

This has some interesting consequences. A 100kg mass has a greater weight at high latitudes than at low latitudes, and so requires greater strength to lift it above your head. An object will weigh about 0.5% more at the poles than at the equator.[4] If you want to break a weightlifting record then head for the equator. The best bet is in Mexico City, where the combination of latitude and altitude means $g = 9.779\text{m/s}^2$ so 100kg weighs 977.9N, and one of the worst venues is Oslo or Helsinki, where $g = 9.819\text{m/s}^2$ and 100kg weighs 981.9N.

66

Googling in the Caribbean

Most sports create league tables to see who is the best team after all the participants have played each other. How the scores are assigned for wins, losses and draws can be crucial for determining who comes out on top. As we have seen, some years ago football leagues decided to give 3 points for a win rather than 2 in the hope that it would encourage more attacking play. But somehow this simple system seems to be rather crude. After all, should you not get more credit for beating a top team than one down at the bottom of the league?

The 2007 cricket World Cup in the Caribbean provides us with a good example. In the second phase of the competition, the top eight teams played each other (actually each had played one of the others already in the first stage and that result was carried forward so they only had to play six more games). They were given 2 points for a win, 1 for a tie, and zero for a defeat. The top four teams in the table went on to qualify for the two semi-final knockout games. In the event of teams being level on points they were separated by their overall run-scoring rate. The table shows the results:

SUPER EIGHT STANDINGS

TEAM	M	W	D	L	Run Rate	Points
Australia	7	7	0	0	2.40	14
Sri Lanka	7	5	0	2	1.48	10
New Zealand	7	5	0	2	0.25	10
South Africa	7	4	0	3	0.31	8
England	7	3	0	4	–0.39	6
West Indies	7	2	0	5	–0.57	4
Bangladesh	7	1	0	6	–1.51	2
Ireland	7	1	0	6	–1.73	2

But, let's think about another way of determining the team rankings that gives more credit for beating a good team than a bad one. We give each team a score that is equal to the sum of the scores of the teams that they beat. Since there were no tied games we don't have to worry about them. The overall scores look like a list of eight equations:

$$
\begin{aligned}
A &= SL + NZ + SA + E + WI + B + I \\
SL &= NZ + WI + E + B + I \\
NZ &= WI + E + B + I + SA \\
SA &= WI + E + SL + I \\
E &= WI + B + I \\
W &= B + I \\
B &= SA \\
I &= B
\end{aligned}
$$

This list can be expressed as a matrix equation for the quantity $\underline{\mathbf{x}}$ = (A, NZ, WI, E, B, SL, I, SA) with the form $\underline{\mathbf{A}}\ \underline{\mathbf{x}} = K\ \underline{\mathbf{x}}$, where K is a constant and **A** is an 8 × 8 matrix of 0s and 1s denoting defeats and wins, respectively, and is given by:

	A	NZ	WI	E	B	SL	I	SA
A	0	1	1	1	1	1	1	1
NZ	0	0	1	1	1	0	1	1
WI	0	0	0	0	1	0	0	0
E	0	0	1	0	1	0	1	0
B	0	0	0	0	0	0	0	1
SL	0	1	1	1	1	0	1	0
I	0	0	1	0	1	0	0	0
SA	0	0	1	1	0	1	1	0

In order to solve the equations and find the total scores of each team, and hence their league ranking under this different point-scoring system, we have to find the eigenvector of the matrix **A** which has all of its entries positive or zero. Each of these solutions will require K to take a specific value. This corresponds to a solution for the list $\underline{\mathbf{x}}$ in which all of the teams have positive scores (or zero if they lost every game), as is obviously required for the situation being described here. Solving the matrix for this so-called 'first-ranked' eigenvector, we find that it is given by:

$$\underline{\mathbf{x}} = (A, NZ, WI, E, B, SL, I, SA)$$
$$= (0.729, 0.375, 0.104, 0.151, 0.153, 0.394, 0.071, 0.332)$$

The ranking of the teams are given by the magnitudes of their scores here, with Australia (A) at the top with 0.729 and the West Indies (WI) at the bottom with 0.071. If we compare this ranking with the table at the top we have:

My ranking	A	SL	NZ	SA	B	E	WI	I
League position	A	SL	NZ	SA	E	WI	B	I

The top four teams qualifying for the semi-finals finished in exactly the same order under both systems, but the bottom four are quite different. Bangladesh only won one game, so scored a mere 2 points and finished second from bottom of the World Cup league. Under our system they jump up to fifth because their one win was against the highly ranked South Africans. England actually won two games, but only against low-ranked teams, and ends up ranked just behind Bangladesh (although it takes the third decimal place to separate them – 0.153 vs 0.151). The poor West Indies finished sixth under the straightforward league system but drop a position under the rank system.

This system of ranking is what lies at the root of the Google search engine. The matrix of results when team i plays team j corresponds to the number of links that exist between topic i and topic j. When you search for a term, a matrix of 'scores' is created by the massive computing power at Google's disposal, which solves the matrix equation to find the eigenvector, and hence the ranked list of 'hits' to the word that you were searching for. It still seems like magic that it happens so quickly, though.

67

The Ice-skating Paradox

When we make choices or cast votes it seems rational to expect that if we first choose K as the best amongst all the alternatives on offer, and then someone comes along and tells us that there is another alternative, Z, which they forgot to include, then our new preferred choice will be to stick with K or to choose Z. Any other choice seems irrational because we would be choosing one of the options we rejected first time around in favour of K. How can the addition of the new option change the ranking of the others?

The requirement that this should not be allowed to happen is so engrained in the minds of most economists and mathematicians that it is generally excluded by fiat in the design of voting systems. Yet, we know that human psychology is rarely entirely rational and there are situations where the irrelevant alternative changes the order of our preferences. A notorious example was the public transport system that offered a red bus service as an alternative to the car. Approximately half of all travellers are soon found to use the red bus. A second bus, blue in colour, is introduced. We would expect one quarter of travellers to use the red bus, one quarter to use the blue bus, and one half to continue travelling by car. Why should they care about the colour of the bus? In fact, what happened was that one third used the red bus, one third the blue bus, and one third their car.

There is one infamous sporting occasion where the effect of irrelevant alternatives was actually built into a judging procedure, with results so bizarre that they eventually led to the abandonment of that judging process. The situation in question was the judging of ice-skating performances at the Winter Olympics in 2002, which saw the young American skater Sarah Hughes defeat Sasha Cohen and favourites Michelle Kwan and Irina Slutskaya. When you watch skating on television, the judges' scores for individual performance (6.0, 5.9 etc.) are announced with a great fanfare. However, curiously, those marks never really determine who wins. They are just used to order the skaters. You might have thought that the judges add up all the marks from the two programmes performed by each individual skater, and the one with the highest score wins the gold medal. Unfortunately, it wasn't like that in 2002 in Salt Lake City.

At the end of the short programme the order of the first four skaters was:

Kwan (0.5), Slutskaya (1.0), Cohen (1.5), Hughes (2.0)

They are automatically given the marks 0.5, 1.0, 1.5 and 2.0 because they have taken the first four positions (the lowest score is best). Notice that all those wonderful individual scores (6.0, 5.9, etc.) are just forgotten. It doesn't matter by how much the leader beats the second-place skater; she only gets a half-point advantage. Then, for the performance of the long programme, the same type of scoring system operates, the only difference being that the marks are doubled, so the first-placed skater is given score 1, second is given 2, third 3, and so on. The scores from the two performances are then added together to give each skater's total. The lowest total wins the gold medal. After Hughes, Kwan and Cohen had skated their long programmes, Hughes was leading, and so had a long-programme score of 1, Kwan was second with a score of 2, and Cohen lay third with a 3. Adding these together, we see that before Slutskaya skated the overall marks were:

Kwan (2.5), Hughes (3.0), Cohen (4.5)

Finally, Slutskaya skated and was placed second in the long programme, so now the final scores awarded for the long programme were

Hughes (1.0), Slutskaya (2.0), Kwan (3.0), Cohen (4.0)

The result is extraordinary: the overall winner was Hughes because the final scores were:

Hughes (3.0), Slutskaya (3.0), Kwan (3.5), Cohen (5.5)

Hughes got placed ahead of Slutskaya because when the total scores were tied the superior performance in the long programme was used to break the tie. But the effect of the poorly constructed rules is clear. The performance of Slutskaya resulted in the positions of Kwan and Hughes being interchanged. Kwan was ahead of Hughes after both of them had skated; but after Slutskaya skated, Kwan found herself behind Hughes! How can the relative merits of Kwan and Hughes depend on the performance of Slutskaya?

68

Throwing the Discus

If you are good at throwing a tennis ball a long way then you might be a good javelin thrower, but you won't necessarily be a good discus thrower. Whereas the javelin relies on a very fast arm and the ability to transfer your forward momentum into the launch speed of the javelin, the discus is about making optimal use of rotational motion whilst remaining confined within a small circle. The world records are old ones: 74.08m by Jürgen Schult in 1986 for the men's 2kg discus, and 76.8m by Gabriele Reinsch in 1988 for the women's 1kg discus. Both athletes were from the old East Germany and their performances have not been improved on since.

Throwing the discus is not like throwing a Frisbee. The thrower is confined to a concrete circle 2.5m in diameter and has to launch the discus within a safety arc of 34.9 degrees. The optimal technique that has developed is to generate fast rotation before releasing the discus at an angle between 30–40 degrees above the horizontal. You will see that the throwers have an initial wind-up phase, swinging back and forth, and then a right-handed thrower will spin in an anticlockwise direction, accelerating to maximum speed at the moment when the discus is released with the arm fully outstretched; it is given some clockwise spin by the first or middle finger of the throwing hand as it is released. The technique needs lots of practice and top throwers tend to be older, with considerable experience of exploiting the wind direction, angle of release and discus weight distribution to best effect. Remarkably, the great American competitor Al Oerter won the men's discus at the 1956,

1960, 1964 and 1968 Games, although he was never the favourite to win on any occasion. His Olympic opponents were increasingly unnerved by his presence over the years.

The one and a half turns that a top thrower makes means that he is accelerating the discus through a distance equal to about one and a half circumferences of the throwing circle, about $1.5 \times \pi \times 2.5\text{m} = 11.8\text{m}$, and will launch it at about $V = 25\text{m/s}$. The inward acceleration needed to sustain circular rotation at this speed is V^2/R where $R = 1\text{m}$ is the length of his arm. This is 625m/s^2, or 63.8g. The world record holder weighs 110kg, so the force on the thrower's arm is over half his body weight. This is why you have to be so strong to throw the discus a long way.

Once the discus is released it behaves like an unpowered aircraft wing. It will be slowed by air resistance but experiences aerodynamic lift due to the air that it displaces as it moves. The air drag and the lift are both proportional to the air density, the cross-sectional area of the discus in the direction of motion (about 0.04m^2), and the square of its velocity relative to the air. The lift will be larger than the drag when the discus has a small angle of attack relative to the air.

The effect of the wind is unusual. We are used to a headwind having an adverse effect on the motion of runners and cyclists but it can be very advantageous for the discus thrower. When the lift force dominates over the drag it will be larger when throwing into a wind because it is proportional to $(V_{disc} - V_{wind})^2$ and so it will be much larger with a headwind (V_{wind} negative) than with a following wind (V_{wind} positive). Detailed studies show that the optimal situation is to throw into a 10m/s headwind, where you will gain about 4m in distance over your performance in still air.[1] By contrast, if you have a 10m/s tailwind behind you it will reduce your throw by about 2m. The figure shows the expected effects and suggests that a following wind of 7.5m/s is the worst-case scenario. These wind speeds are very strong and would invalidate sprint performances where the maximum allowed following wind is 2m/s for record purposes. It is strange that these major wind effects are not taken into account in the keeping of discus records[2] – they are often more important than illegal following winds in the sprints and jumps.

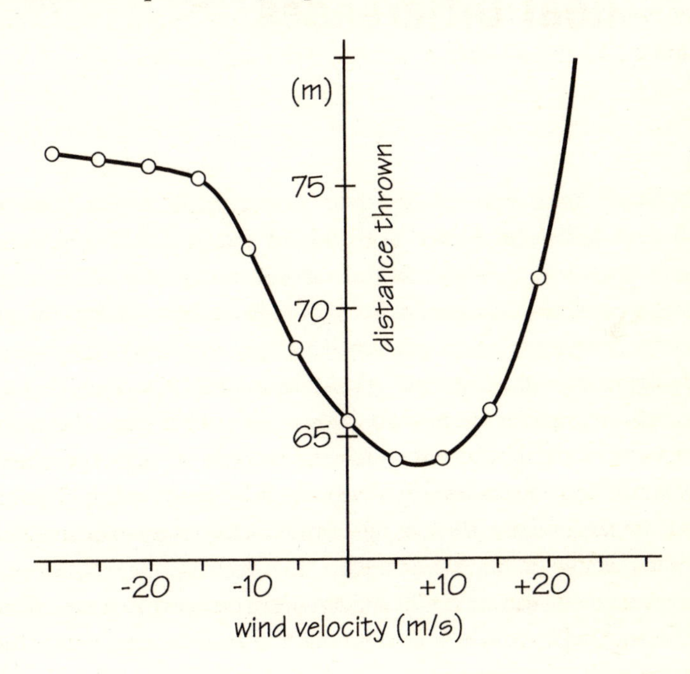

69

Goal Differences

Different sports have devised their own solutions to the problem of ties. Americans have a deep fear of drawn games and there must always be a winner in their sports competitions, but other nations are more sanguine. Yet although about a quarter of all soccer matches end in draws, when it comes to deciding the league championship or a cup final, it's different: there's got to be a way to split teams that are tied on goals or on points. Over the years all sorts of methods have been tried to determine the outcome of a cup final. Extra time is always the first resort and if it failed to produce a winner then, in the old days, there used to be a replay – and possibly even a second replay. More recently, there have been experiments with varieties of sudden-death extra time: the so-called 'Golden Goal' – in which the first team to score in extra time wins the match – has been used in hockey, ice hockey and soccer. It was a short-lived experiment in soccer though (from 2002–4) and was superseded for a while by the 'Silver Goal' in which the team leading after the first fifteen-minute half of extra time would win the match and the second period of extra time wouldn't be played. Otherwise, it was the team that led at the end of the second half, as usual.

In 1967 the European Inter-Cities Fairs Cup quarter final[1] between Glasgow Rangers and Real Zaragoza was decided by a toss of a coloured disc after it finished with the aggregate score drawn at 2–2. In later years these ties were broken by away goals and penalty shoot-outs. If scores were tied on aggregate after two legs then any away goals scored were counted as double. Alas, this caused problems for mathematically challenged referees. Glasgow Rangers (why is it always them?) lost a penalty shoot-out against Sporting Lisbon after winning 3–2 at home and losing 3–4 away. The referee failed to realise that Rangers had won on away goals and there shouldn't have been a penalty shoot-out at all. The Rangers manager, Willie Waddell, tracked down the UEFA official in the ground with a copy of the rule book and the result was overturned in Rangers' favour; they went on to win the Fairs Cup, beating Moscow Dynamo 3–2 in the final.

Deciding ties on points in leagues offers many possible formulae. Once upon a time 'goal average' was used. This was calculated as the ratio of a team's goals scored for (F) to goals conceded against (A):

$$\text{Goal average} = F/A$$

In the days before electronic calculators, journalists would argue that slide rules were needed to determine the ordering of the league. Goal average was replaced, first at the 1970 World Cup Finals, and then in the 1976–7 English Football League season, by goal difference:

$$\text{Goal difference} = F - A$$

In 1952–3, the English League Championship was decided on goal average after Arsenal and Preston North End both amassed 54 points (only 2 points for a win then) but Arsenal had $F/A = 97/64 = 1.5156$ whereas Preston's goal average was only $F/A = 85/60 = 1.4167$. Goal difference was introduced in order to encourage attacking play – rewarding large F more than small A. In 1952–3, we see that Arsenal had a goal difference of 33 while Preston had 25. It is very unlikely that two teams will end up with the same goal average but there are a lot more ways for goal differences to be the same. Remarkably, in the 1988–9 season, Arsenal and Liverpool finished equal on 76 points, and also on goal difference, at the top of the Football League Championship. In fact, they had identical records: won 22, drawn 10 and lost 6. Arsenal had a goal difference of 73–36 = 37 and Liverpool had 65–28 = 37. If goal average had still been used Liverpool would have won the league, with F/A = 65/28 = 2.3 as opposed to Arsenal's F/A = 73/36 = 2.0.

Fortunately, in 1989 the league had a rule in place to decide this tied goal difference and the championship went to Arsenal because they had scored more goals (i.e. bigger F) – 73 compared to Liverpool's 65. This tie-breaking rule was abandoned after that and was not adopted by the new Premier League when it formed in 1992.

Football leagues still use goal difference to resolve tied points, but is it the best scheme? I can think of many alternatives. Why bother with points at all? Instead, just determine all the league positions by goal difference. One option would be to award no points for a 0–0 draw, or perhaps whenever there is a drawn game award no points to the home team and 1 to the away team. Here is the outcome of the 2009–10 English Premier League table showing points gained,* goal difference and points totals if there are no points awarded for a home draw:

Team	Total points	Goal difference	Points total with no points for a home draw
Chelsea	86	71	85
Man Utd	85	58	84
Arsenal	75	42	73
Spurs	70	26	65
Man City	67	28	63
Aston Villa	64	13	56
Liverpool	63	26	60

*I have not included Portsmouth's 9-point deduction for financial insolvency.

Everton	61	11	55
Birmingham	50	–9	41
Blackburn	50	–14	44
Stoke	47	–14	41
Fulham	46	–7	43
Sunderland	44	–8	37
Bolton	39	–25	33
Wolves	38	–24	32
Wigan	36	–42	29
West Ham	35	–19	30
Burnley	30	–40	25
Hull	30	–41	24
Portsmouth	28	–32	25

As you can clearly see, almost nothing is gained at the top by all the detail of awarding different points for wins and draws. Near the bottom of the table there are more interesting variations and Wigan would have been relegated (their goal difference was hit by 9–1 and 8–0 drubbings by Spurs and Chelsea respectively). The new points totals calculated with no credit for home draws is interesting. It makes no difference at the top and the bottom but creates several changes in position in the mid-table region. Although it has a big impact on the points totals of teams like Aston Villa, Birmingham and Blackburn who drew lots of home games, it didn't alter their positions because of the huge points gaps between groups of teams in the table.

Not surprisingly, as winning is about scoring more goals than you concede, goal difference is very closely correlated with points awarded. Thus a league table based on goals scored would be simpler and offer a greater incentive for goal scoring, After all, this is what fans (and TV companies) want to see.

70

Is the Premier League Random?

Football looks pretty random at times, especially if you watch the lower divisions. Players kicking the ball anywhere in desperation, quick sequences of passes and interceptions, rebounds, own goals; you begin to wonder whether this all averages out over a season to produce a league table that displays many features of a simple random process. On average, about one football match in four is a draw. Let's assume that the Premier League is a simple random process in which any game has a 1 in 4 chance of being a draw. For simplicity, we will ignore home advantage and assume that there is a 3/8 chance that a game is a home win and a 3/8 chance that it is an away win. There are twenty teams in the league and each of them plays the others twice, home and away, so they each play thirty-eight games.

We can play our random league between the twenty teams using an eight-sided spinner with two sides labelled 'draw', three labelled 'home win' and the three others 'away win'. But you would need a lot of time to play all the games by hand like this. Better to ask a computer to do it for you in a few seconds. Now total up the points gained (3 points for a win, 1 for a draw and zero for a defeat) by each of the teams you named and create the final league table.[1] The teams with the most points we call team 1, the next team 2, and so on.

For comparison, I have shown the actual final Premier League tables for the seasons 2003–4 and 2004–5. A comparison between the random league and the actual results is quite instructive.[2]

Team	Won	Lost	Drawn	Points	2003–4 actual league table	2004–5 actual league table
1	19	10	9	67	90 Arsenal	95 Chelsea
2	18	9	11	63	79	83
3	18	8	12	62	75	77
4	17	10	11	61	60	61
5	16	10	12	58	56	58
6	16	8	14	56	56	58
7	13	16	9	55	53	55
8	15	9	14	54	53	52
9	16	5	17	53	52	52
10	15	8	15	53	50	47
11	15	8	15	53	48	46
12	14	11	13	53	47	45
13	13	13	12	52	45	44
14	14	9	15	51	45	44
15	13	12	13	51	44	42
16	15	4	19	49	41	39
17	11	11	16	44	39	34
18	9	16	13	43	33	33
19	9	8	21	35	33	33
20	8	7	23	31	33 Wolves	32 Southampton

You notice that if you ignore the top three teams the rest of the league table looks very similar to a random distribution of points outcomes. The top three teams are quite different because they have a much greater chance of winning their games than the 3/8 (= 37.5%) model we adopted. In fact, the winning Chelsea team won 76% of their games and the winning Arsenal team more than 68% of theirs. In both seasons the Premiership was so uncompetitive that the fourth-place team was closer to being relegated than to winning the league! One of the beauties of our model is that the identity of the winning team is random and this adds interest to the competition by combining aspects of the Premiership with those of the National Lottery.

Look closely and you will see that after ignoring the top three teams the random model works very well near the top and near the bottom but drifts away from the actual results in mid-table. There is a good reason for this. Our simple model has assumed an equal probability for a home or an away win. This is a good approximation for the teams at the top – they win almost all the time, home or away – and for the teams at the bottom – they lose most of the time, whether they are home or away. In mid-table, a home advantage is usually advantageous and these teams will win more often at home than away. This defect could be repaired by slightly changing our model, say with a quarter of games being draws still but 7/16 home wins and 5/16 away wins. It would also be improved more by inserting a few 'super teams' who have a 3/5 chance of winning their games.

Overall, this simple model should offer encouragement to all those who don't like watching football. The whole outrageously expensive business of the Premier League (and the other divisions too) could be replaced by a simple random number generator without too much change to the overall pattern of the results – and it might be Arsenal's best chance of winning the league again in the future.

71

Fancy Kit – Does It Help?

Athletes in speed events like sprinting, cycling and speed skating have a history of experimenting with bodysuits and hoods to cut down on air resistance. When female athletes like 'Flo-Jo' started appearing on the track in one-legged bodysuits you began to wonder whether there was any athletic advantage to be gained at all from these fashion statements, generously funded by sponsors hoping to sell them en masse to lesser mortals. Also puzzling is the trend that suddenly emerged for female runners to wear crop tops which left their midriffs exposed to the elements. No male athlete wears a half-length running vest. There is a good reason for this; if it's cold you get cold because of skin exposure; if it's quite warm you get cold as well because of evaporation of sweat; and if it's very sunny your skin is unduly exposed to the sun. All in all, running in a crop top makes little sense for males or females.

The motivation for bodysuits was to reduce air resistance. This is worth doing. If you run at speed V relative to the ground into an oncoming wind with speed W relative to the ground, then you will feel a resisting (i.e. negative) drag force due to the air equal to:

$$F = -\tfrac{1}{2} C\rho A(V-W)^2$$

where ρ is the density of the air you are running through, A is the frontal cross-sectional area of body and kit, and C is the so-called drag factor that depends on your body shape and the aerodynamic quality of your surface.[1] Its value is typically quite close to 1 for runners and about 0.8–0.9 for a racing cyclist.

There are a few things to notice about this formula. It depends on the square of the runner's speed relative to the wind. The runner's speed, V, is always positive running forwards and the speed of the wind, W, will be positive if it is following you but negative if it is a headwind blowing into your face. Sprint races and jumps record a following wind by a + sign and a headwind by a – sign.[2]

The factors under an athlete's partial control in the drag-force formula are the area A and the drag factor C. You can reduce the drag by reducing the area that you present to the air. Waving your arms about and having loose-fitting flappy clothing is going to increase A and slow you down. Typically, runners will present to the air a frontal area of about $A = 0.45m^2$, and C is about 1, so a world-class male sprinter moving at 10m/s in windless conditions ($W = 0$) will expend about 3% of his effort in overcoming wind resistance.

You can reduce A very slightly by adopting different postures, and some runners will respond to the wind conditions in this way. Is long hair going to be a problem? Dangling dreadlocks will certainly increase A, and the accompanying drag, but the effect is small. The head only occupies 6–7% of the frontal body area and this fraction of the overall 3% effort against the wind is the most that could be attributed to the whole head. Hair will be a still smaller fraction of this. The hoods that were worn by some sprinters in the late 1980s are therefore unlikely to offer any significant advantage and probably make you feel uncomfortably hot.

What about bodysuits? They aim to reduce the drag factor, C, created by the sprinter's moving body. Interestingly, the drag factor changes quite significantly for small changes in speed around a critical speed of flow and this is studied very carefully in wind tunnels for all sorts of moving bodies, especially cars and planes. The sudden change is caused by the airflow becoming turbulent very close to the surface of the moving body and is called the 'drag paradox' by aerodynamicists. This turbulence can be induced by the roughness of the surface over which the air is flowing and so it is best for runners to wear smooth, tight-fitting kit with very thin ribbing, about 0.5mm in height to trip the smooth flow and make it turbulent very close to the surface. If done correctly, this can halve the value of C and improve a 100m sprinter's time by a crucial 0.1s. However, at the slower speeds found in the parents' race at the school sports you will not be in the speed range that benefits from the sudden reduction in drag provided by the 'drag paradox' and Lycra bodysuits are not recommended.

72

Triangles in the Water

Water polo is a mixture of swimming, handball and wrestling. Its name derives from the Victorian British–Indian usage of the word 'polo', to mean a 'ball', and has nothing to do with horses. Men's water polo has been in the Olympic Games since 1900, but a women's tournament was added only in 2000. Teams of seven swimmers, one of whom is a goalkeeper, play four exhausting eight-minute periods and the clock stops when the ball is not in play so the periods tend to last about twelve minutes in total. Only the goalkeeper can put his feet on the pool bottom, although the water depth must be at least 1.8m (6ft) in international competition and there is no shallow end. You are treading water ('eggbeating' in US parlance) or swimming fast all the time. I used to play water polo at school and can confirm that it is exhausting – you swim a long way during those periods, constantly accelerating and turning, pushed and pulled by opposing players trying to drown you! You also want to avoid getting hit by the heavy ball. And once you are in possession of the ball you have thirty seconds in which to make a shot on the opponent's goal or possession passes to them. Only the goalkeepers can touch the ball with both hands at once and you are not allowed to submerge the ball.

Very quick thinking time and eye–hand coordination is required for long periods in water polo. There is a lot of physical contact and fouls are frequent. Like in ice hockey, a player can be 'sin-binned' for twenty seconds (or until his team regains possession, or a goal is scored) or removed entirely if they commit three sin-binnable offences (although another player can come on as a substitute after four minutes of banishment). These penalty periods for foul play sound very short but they often occur, and with only six outpool players in a team the effect of being one player down is proportionately huge. It is very hard to avoid conceding a goal in those circumstances and, unlike football, there is no scope for time wasting. Even if you avoid conceding the goal, you will have expended a lot of extra effort defending and may pay the price later.

When you have a temporary one-player advantage there are clear geometric strategies for pressing it home and turning it into a goal. Water polo defenders try to get between pairs of attackers and cover them both so you want to make that difficult to do. Also, most obviously, you can throw the ball through the air much faster than any goalkeeper can paddle sideways through the water. The goal is 3m wide and extends 90cm above the water. Attackers therefore try to create long and short triangles for passing movements across the pool that will leave the goalkeeper floundering on the wrong side of the goal and unable to reach the incoming shot from the other side. Attackers try to avoid making horizontal passes to players who are exactly in line with them because they are hard to receive and there is lots of scope for error. Instead they look to make diagonal passes to players who are at different distances from the goal than themselves. An expert player can receive and then shoot the ball towards goal in a single movement without allowing the ball to touch the water. Sometimes the final shot will be entirely in the air but another option is to fire it towards the water just in front of the goalkeeper so that it 'skims' fast off the surface and is harder for the goalkeeper to read.

A typical attacking formation when attackers have a six players to five advantage over the defenders is the 4–2 configuration; with a line of four nearest the goal and another of two a few metres back. You can't receive the ball closer than 2m (the red zone) from the goal, so you can't be that close to the goal unless the ball is as well.

Goal

A1 A2 A3 A4

A5 A6

This is much better than two lines of three because in the latter the defenders can sit in between A1 and A3 and between A4 and A6 and cover two players each. If you do want to play like that then you need to have A5 drop back to create a third line. This player will not be covered and is always free to take a shot at the goal. This opens up lots more diagonals for passing and makes the defender's job much harder.

<table>
<tr><td></td><td>Goal</td><td></td><td></td><td></td><td>Goal</td><td></td></tr>
<tr><td>A1</td><td>A2</td><td>A3</td><td></td><td>A1</td><td>A2</td><td>A3</td></tr>
<tr><td></td><td></td><td></td><td>or much better . . .</td><td></td><td></td><td></td></tr>
<tr><td>A4</td><td>A5</td><td>A6</td><td></td><td>A4</td><td></td><td>A6</td></tr>
<tr><td></td><td></td><td></td><td></td><td></td><td>A5</td><td></td></tr>
</table>

Looking at the 4–2 configuration, we can see the geometrical possibilities. There are nice big triangles to work the ball around, A1–A6–A4 or A1–A5–A4. If you can draw the goalkeeper to the same side of the goal as attacker A1, then a very fast triangular movement to A4 via A6 or A5 will move the ball faster than the goalkeeper can back-paddle to cover the incoming shot. Another triangle is created by the A1–A5–A6 trio. In all cases you can see that the key passes are diagonals of the triangle and awkward sideways passes are avoided.

You don't have long to take advantage of an extra player so these triangles need to be practised carefully and at high speed. There is a lot of tactical expertise needed in water polo – it is as if corners and free kicks in soccer were happening constantly. There is not much the defenders can do to counter the extra player except have a very big, fast-moving goalkeeper. So, if you get a chance to watch top-flight water polo at the Olympics, take it.

73

The Illusion of Floating

Top basketball players, and even footballers, will sometimes appear to be hanging in the air when they jump to guide the basketball through the ring or head a football towards goal. This is strange because the laws of mechanics don't allow the motion of a projectile to spend a period 'hanging' in the air before continuing to fall under the force of gravity to the ground. This leads many people to believe that these reports of superhuman sporting movement are pure illusion, or hyperbole, created by overenthusiastic sports fans and crazed commentators.

The sceptic says that when a projectile, in this case the human body, is launched from the ground then its centre of mass (about 0.55 of your height) will follow a parabolic trajectory and nothing the projectile can do will change that. However, there is some fine print to the laws of mechanics: it is only the *centre of mass* of the projectile that must follow a parabolic trajectory. If you move your arms around, or tuck your knees into your chest, you can change the location of parts of your body relative to your centre of mass. Throw an asymmetrical object, like a screwdriver, through the air and you will see that one end of the screwdriver may follow a rather complicated backward looping path through the air. The centre of mass of the screwdriver, however, follows the same old parabolic trajectory regardless.

Now we begin to see what a basketball player can do. His centre of mass follows a parabolic trajectory but his head doesn't need to. He can change the shape of his body so that the trajectory followed by his head stays at one height for a noticeable period, up to half a second. When we watch him jumping we only notice what the head is doing and don't watch the centre of mass. Michael Jordan's head really does follow a horizontal trajectory for a short time. It's not an illusion and it doesn't violate the laws of physics.

This trick is also seen more beautifully executed by ballerinas who perform the spectacular ballet jump called the *grand jeté* in which they launch themselves into the air and perform a full leg split. They try to create the illusion of floating in the air for artistic reasons. During the jump phase they raise their legs up to the horizontal position and put their arms above their shoulders. This raises the position of their centre of mass relative to their head. Soon afterwards, their centre of mass falls relative to their head as the legs and arms are lowered downwards during their fall back to the floor. The ballerina's head is seen to move horizontally in mid flight because her centre of gravity is shifting up her body during the jump phase. Her centre of mass follows the expected parabolic path throughout, but her head maintains the same height above the floor for about 0.4s to create a wonderful illusion of floating. Physicists have monitored the motion of dancers with sensors and the figure below shows the variation of the displacement of a dancer's head from the ground during this jump. There is a very distinctive plateau in the middle of the jump that shows the hang in the air and is quite different to the parabolic path followed by her centre of mass.

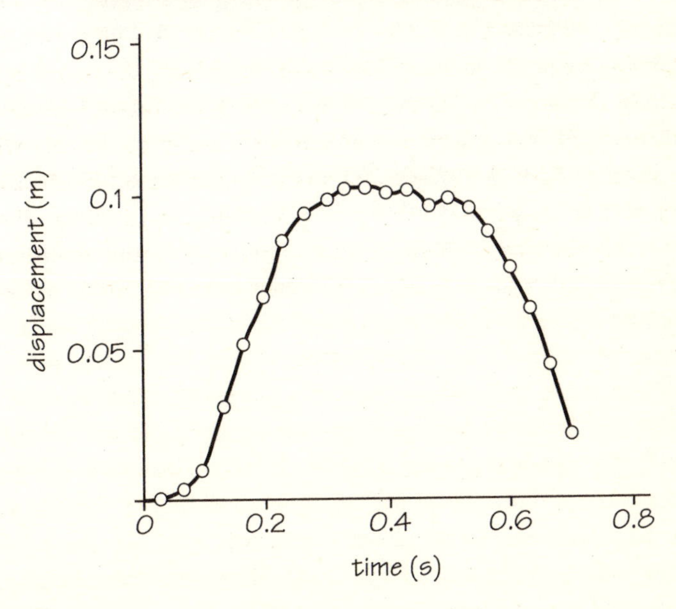
0.15
0.1
0.05
displacement (m)
0
0.2
0.4
0.6
0.8
time (s)

74

The Anti-Matthew Effect

In many sports, notably football in England, Italy and Spain, there is a wealth trap. The rich clubs get richer, buy more players, win more trophies, get more TV revenues, get richer still, win more trophies, and so on. At the moment, UEFA is introducing a form of means testing for clubs that determines their income from football-based activity alone. However, there seem to be so many caveats, safety nets and transition periods to prevent anything punitive happening to any of the top clubs that it remains unclear whether these measures will have any real effect.

The American National Football League (NFL) plays another sort of football but generates an awful lot of money, just like the Premier League in England. Yet unlike in European football, the NFL has been careful to avoid money breeding irreversible success with rich teams becoming richer and more successful so that no one else will ever win their championship. They have instead created a counter to runaway domination. Each spring, the NFL organises a 'draft' in which the clubs recruit new young players who have not been out of high school for more than two and a half years. Prospective players sign with an agent and join the draft pool. The interesting thing about the selection process is that the least successful team from the previous season is given first choice of the players, and so on, all the way down to the last choice that goes to the previous year's Super Bowl champions. There are seven rounds of this selection process, with thirty-two selections in each round, run over a few days near Easter each year. Drafted players are paid salaries that reflect the order in which they are selected with the earliest picks getting the largest contracts.

This process seems to be quite successful in maintaining the competitiveness of the NFL. If you were the best last season then your rivals are going to have an advantage over you in this year's draft and the gap between the teams will be kept close.

This type of situation can be modelled using an interesting piece of mathematics called a delay-differential equation. If the success (income, games won or trophies acquired) of a team at time t is denoted by S(t) then its rate of change might be assumed to be proportional to the amount of success at present:

$$dS(t)/dt = FS(t)$$

This equation has solutions of the form $S(t) = A\exp(Ft)$ where A is a constant. So, if F is a positive constant quantity, the success grows exponentially with time (which is rather like the top of the Premier League); but if F is negative it decreases exponentially with time. The NFL draft is tantamount to changing this equation into one where the rate of increase in S at time t is proportional, not to the value of S at time t, but to its value at some past time t–T, which we call S(t–T), where T is a constant number – the 'delay'. Now, we have an equation:

$$dS(t)/dt = FS(t-T)$$

that is subtly different in its behaviour. It has steady oscillating solutions that vary as:

$$S(t) = A\cos\{\pi t/2T\}$$

over time, where A is a constant determined by the starting conditions. These oscillations have a period[1] from peak to peak that is equal to 4T.

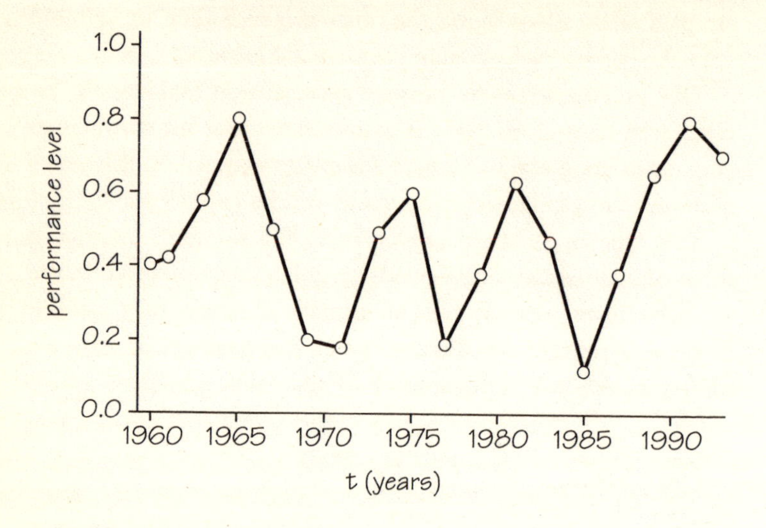

Does it work? The records of the teams in the NFL seem to follow this type of oscillating behaviour over a period of years. They are using a delay model with T = 2 years approximately and so we expect to see a 4T = 8 year periodic cycling of every team's fortunes from the effect of the draft alone. Robert Banks studied the performance of the Buffalo Bills and the Chicago Bears (by scoring 1 point for a win, 0.5 for a draw and zero for a loss) and they show a clear average 8.3 year and 8.0 year cycle of success respectively.[2] The Buffalo Bills performance is shown in the figure above. Averaging across the whole NFL, he found an 8.24 year success cycle which fits this simple delay equation model quite well. In reality, there are other factors at work which influence success – good management, good luck, injuries, well-chosen tactics – but the success of this model suggests that the NFL draft's ability to ensure bad teams get better and the best teams get worse is a simple delay system. It would be a positive innovation in European football. All the players seeking to be transferred, or entering the game from youth teams, would be named at the start of the season and offered first to the least successful teams in one league.

75

Seeding Tournaments

In knockout tournaments, like the Wimbledon Tennis Championships or the Champions League, there is a seeding of entrants to ensure that there is a good chance of the best entrants avoiding each other until the later stages of the competition. In the case of Wimbledon, the seeding is highly structured according to the current world rankings which are created from past tournament results. The top two seeds are at the opposite ends of the draw and if each wins all their games then they will meet in the final. In football's Champions League there are added constraints at work to avoid teams who met in the (pointless) group stages from meeting again early in the knockout phase. The English FA Cup is a purer competition that sees the two lower league clubs enter the competition in the first round after the five qualifying rounds are completed by non-league teams. The Premier League and Championship League teams enter in the third round, but the third-round draw is completely random: the top two teams could meet each other straight away. Each subsequent round is redrawn at random so the teams cannot foresee who they may (or cannot) play in future rounds. This is the most natural and exciting structure to use for a knockout tournament. Top teams can be drawn away against minor 'giant killers' and there is a strong element of unpredictability. Of course, the top clubs hate it. They want money out of the competition and this is why the Champions League has innumerable tedious games in the qualifying league sections to determine who is in the knockout phase. It ensures that no one can be eliminated after two (home and away) games. In fact, even if you are knocked out in the qualifying leagues, you get placed in the next round of the second-string Europa League. It's like those *Strictly Come Dancing* shows on TV that take forever to eliminate any competitor.

If we have N competitors in a knockout cup then we will need r rounds, where $N = 2^r$. The FA Cup has $256 = 2^7$ teams and so there are seven rounds before the final. The tournament organisers have made sure that the earlier qualifying rounds give rise to a number of teams in the first round proper that is an exact power of two. What if your tournament could have any number of entrants? In that case you need to give $2^r - N$ of the entrants a bye in the first round. Then, the number left in the next round will be a power of two, and everything goes through as before. For example, if you have twenty-eight entrants you need to give $2^5 - 28 = 32 - 28 = 4$ of them a first-round bye. The other twenty-four will play each other, and the twelve winners plus the four byes will give sixteen teams for the second round.

In a tournament where the whole future patterns of who plays who is (unlike the FA Cup) completely defined initially, we can work out the chance that the two best teams in a competition with $N = 2^r$ entrants will meet in the final. We note that if they are both drawn in the same half of the draw this certainly can't happen. The best they could hope for is to meet in the semi-final. There are ½N teams in each half of the draw. After choosing the best team's place, there are ½N–1 places in the same half of the draw and ½N in the other half of the draw. Since they can only meet in the final if they are in opposite halves of the draw, we see that this chance is $(\frac{1}{2}N - 1)/(N - 1)$. So, if there are N= 32 entrants, the chance of the best two meeting in the final is 15/31, which is a bit less than 50%. As the number of entrants gets larger, so the chance of the best two meeting in the final gets closer and closer to 1 in 2.

76

Fixing Tournaments

We have looked at some of the basic features of structuring and seeding knockout tournaments. It is clear that if the tournament's development is completely defined by the draw for the first round then you can avoid stronger players early on. Also, some stronger players may get eliminated because they meet even stronger opponents at the outset. Could you so arrange the draw that a relatively weak player would win the tournament?

To achieve this, we need to reverse-engineer the event. For simplicity, let's have eight entrants and call them A, B, C, D, E, F, G and H. What needs to happen for H to win the championship? There will be two rounds before the final so if everyone plays to form then H has to be better than at least three other players.

H bts G, E bts F, D bts C, A bts B
H bts E and A bts D
H bts A

The most unusual outcome we could engineer would be for H to be better than only the three opponents, A, E and G, who she needs to beat to win the title. Next, we need A to be better than B, and D and E to be better than F. We see that H does not need to be ranked very highly overall – just fifth out of the eight competitors – and she would lose to F, B, D or C if she was drawn against them but she can still win the championship if the initial draw is manipulated in her favour.

77

Wind-assisted Marathons

On 18 April 2011, the winner of the Boston marathon lopped nearly a minute off the world record for the men's marathon. The Kenyan runner Geoffrey Mutai, who is no relation of his namesake Emmanuel Mutai who won the London marathon the day before, completed the point-to-point marathon in 2hr 3m 2s. This was 57s faster than Haile Gebrselassie's world record – a huge advance. Remarkably, the second-placed athlete, Moses Mosop, was only 4s behind.

Runners like point-to-point marathons. There are no boring laps, you're not going to lap a troop of tail-enders wearing Superman outfits and giraffe suits late in the race – and you feel that you are getting somewhere. Spectators, organisers and the media, by contrast, much prefer to have the runners doing many small laps around a single loop course. There are more opportunities to see the passing runners, you need fewer feeding and water stations, and you can more easily monitor what's happening.

There are other potential differences between point-to-point and loop marathons. The point-to-point race could be downhill, for instance! To avoid this advantage, a record can only be set in a marathon race where the finish is no more than 42m lower than the elevation of the start, regardless of how hilly things are in between. Unfortunately, the elevation drop on the Boston course is very advantageous and well over this legal limit – 139m – although it is ameliorated by the hills along the way.*

The elevation drop was not the most controversial aspect of the Boston point-to-point world 'record' in 2011. If the wind is blowing in the right direction in a point-to-point you can end up running a wind-assisted marathon for the whole 26 miles – remember that times or distances in track sprints, hurdles and horizontal jumps are invalid for record purposes if the following wind exceeds just 2m/s. Winds were significant in Boston, around 15mph or 6.75m/s according to some commentators. The strong following wind at the event is also attested to by the fact that the press reported many runners afterwards saying they could feel no wind over large parts of the course. This is exactly what you would feel if your forward running speed was equal to the following wind speed – and the winner's average speed was 5.7m/s, rather similar to the anecdotal wind-speed estimates.

The drag force opposing your forward motion when you are running at speed V with a tailwind of speed W is:

$$F_{drag} = -\tfrac{1}{2}\,\rho CA(V - W)^2$$

where C is the drag factor, A is your body's cross-sectional area into the wind, and $\rho = 1.2\text{kg/m}^3$ is the air density. For typical elite marathon runners (Geoffrey Mutai has a mass of about 60kg and height 1.83m) the quantity CA is about 0.45m^2. You can see from this formula that as the wind speed nears your running speed, $W \to V$, you are in effect running without having to work ($F_{drag} \times V$) against the air drag at all.

Mutai's average speed in his 'record' Boston run was 42,195/7,382 = 5.7m/s. This is about 13mph. The average drag force in still air

*The Tucson marathon descends more than 600m from start to finish. This is like being pushed along by an average gravity force equal to 0.014 of your body weight – 8.2N for a runner of mass 60kg.

($W = 0$) running at 5m/s is $0.5 \times 1.2 \times 0.45 \times 25 = 6.8$J/m. Studies of elite runners show that running on the flat they can maintain energy consumption at a level of about 3.6J/kg/m, largely independent of their speed in the racing range. With a mass of 60kg, this would mean he was requiring $3.6 \times 60 = 216$J/m to run in a steady state in still air. The strong following wind saved him about 6.8J/m (not far off the purely gravitational advantage in the downhill Tucson marathon) and so the ratio of the power needed running with tailwind to that without is $(216–6.8)/216 = 0.97$ – that is, a 3% saving. Since the power expended is proportional to the speed he achieves, and his finishing time will be inversely proportional to this speed, we see that the effect of the tailwind will be to reduce his finishing time of 7,382s by about $7{,}382 \times 0.03 =$ 222s, which is 3m 42s. This makes Mutai's wind-aided performance equivalent to a winless marathon time of about 2hr 6m 45s.

However, one final factor that might upgrade Mutai's performance is to allow for the fact that the leaders ran in a closely bunched group for the first half of the race and he may not have felt the full benefit of the wind at his back. If we were to assume that he gained advantage from the wind only in the second half of the race, which was run in an amazing 61m 4s – this time alone suggests that he was helped more by the wind – then there would have been an overall advantage of 110s in that part of the race, and the total time would have been effectively reduced to about 2hr 4m 52s. The real benefit of the wind probably therefore lies between 110–222s. Meanwhile, the officially accepted record has advanced to the time 2hr 3m 38s set by Patrick Makan on 25 September 2011.

78

Going Uphill

Runners and cyclists know the effects of hills only too well. Runners lose far more time going uphill than they gain coming down. The cyclists have it better because they can make their downhill descent without making any effort at all but the runner's descent always requires some leg-jolting effort.

A cyclist who has body plus bike mass of 75kg and who cycles at 10m/s will have to overcome a frictional force of about $0.004 \times 75\text{kg} \times 9.8\text{m/s}^2 = 2.94\text{N}$. The power needed to counter this force is obtained by multiplying it by the speed, 10m/s, which gives a power of 29W to overcome the friction of the rolling wheels. More generally, if there is no wind and the cyclist rides at speed V in air of density ρ while presenting a body area A to the oncoming air, with an aerodynamic drag factor C such that $CA = 0.25\text{m}^2$, then he will also have to overcome the air-drag force of $0.5\rho CAV^2 = 0.5 \times 1.2 \times 0.25 \times 100 = 15\text{N}$, and the power needed to do so is therefore $15\text{N} \times 10\text{m/s}^2 = 150\text{W}$.

We see that about five times more work is required to overcome the aerodynamic drag than the friction. The total power needed to be supplied by the cyclist is about $150 + 29 = 179\text{W}$ – nearly enough to power three 60W light bulbs if you rigged our cyclist to a dynamo.

Now, let's compare this effort with ascending up a 1 in G gradient for the first 5,000m and freewheeling down the same gradient for the second 5,000m. When the cyclist goes uphill an extra force has to be overcome – gravity. This force is equal to the weight of the cyclist times the fractional gradient 1/G.[1] In our example, the gravity force is $(75 \times 9.8)/G = 735/GN$. We can see that if the hill is 1 in 10 then this force is 73.5N and is much bigger than the air-drag force.

What happens on the downhill? Now the gravitational force down the gradient works in your favour and accelerates the rider, while the air-drag force and the (much smaller) friction oppose it. When the two forces become equal, the cyclist descends at constant speed because no net force acts. In this situation:

$$0.5\rho CAV_d^2 = mg/G$$

and the constant descent speed is:

$$V_d = \sqrt{(2mg/\rho CAG)} = \sqrt{(2 \times 75 \times 9.8)}/\sqrt{(1.2 \times 0.25 \times G)} = 70/\sqrt{G}m/s$$

For our 1 in 10 gradient, with G = 10, we have a formidable descent speed of 22m/s.

There is a nice connection between the three speeds our cyclist will achieve on the level, V_L, going uphill, V_{up}, and descending, V_d. If he can generate a steady power (power = force × velocity), P, for a significant period when cycling on the level, then $P = 0.5\rho CAV_L^3$. When he goes uphill, this same power output will allow him to overcome the gradient of the gravitational force, so we will have $P = mgV_{up}/G$. Finally, on the descent we have $V_d^2 = mg/(0.5\rho CAG)$. If we combine these equations we have a simple relationship between the three speeds:

$$V_{up} = V_L^3/V_d^2.$$

If he was going at 10m/s on the level then he will descend the 1 in 10 gradient at 22m/s after climbing it at 0.2G = 2m/s.

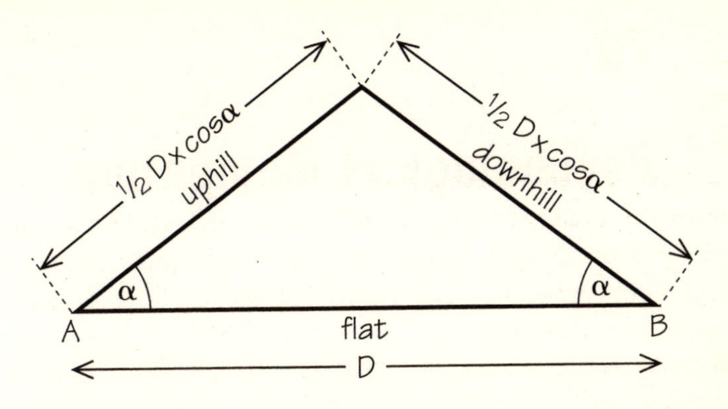

We can also compare the time, T_L, taken by a cyclist to go some distance D from point A to point B all on the level at speed V_L with the time, T_{hill}, it would take the same cyclist to cycle a distance ½DcosA from A up a 1 in G gradient[2] at speed V_{up} and then descend down the other side at speed V_d to arrive at B. For our cyclist, who could maintain 10m/s on the flat, the hilly route would have taken him 2.5 times longer.[3]

79

Psychological Momentum

In many sports encounters there are 'inexplicable collapses', 'inspired performances', 'runs of bad luck', competitors 'lifting their game' or succeeding 'against the run of play', and players coming back to win 'against the odds'. All these phrases, and the unexpected events they describe, suggest that there might be a form of 'psychological momentum' in sport that goes against the cold mathematical assumption that the chances of winning a point or a game is independent of what went before.

If we forget about psychology for a moment, we can analyse a game like tennis by assuming that each player has the same probability, p, of winning a point and the same probability, 1–p, of losing a point, if we ignore the advantages of serving. How does this carry over into the probability of winning a set?

There are several routes to winning the set. The simplest is a straight win in four points from 40–0. It has a probability of p^4 if all the points are independent and neither player is swayed by what happened in the past. For simplicity, we don't use a tiebreak. The probability of winning from 40–15 is $4p^4(1-p)$ because there are four different ways to get to 40–15. Each requires you to win three points with probability p and lose one with probability 1–p and then win the next point with probability p to win the game. The probability of winning from 40–30 needs to count the ten different ways of getting to 40–30 and multiply by the probability of winning the next point, which gives $10p^4(1 - p)^2$. Lastly, we need to work out the probability of winning from deuce. The chance of being at deuce (40–40) counts the twenty different ways of getting there and is $Q = 20p^3(1-p)^3$. What's the chance of winning from deuce? You can either win two straight points, with probability p^2, or lose one then win one or vice versa to bring you back to deuce from which again you have the probability Q of winning. Therefore we have $Q = p^2 + 2p(1-p)Q$ and so $Q = p^2/\{p^2 + (1-p)^2\}$.

We now just add up the probabilities for the different routes to winning the game – 40–0, 40–15, 40–30 and from deuce – to give the overall probability, G, of winning the game in terms of the probability of winning a point, p:[1]

$$G = p^4 + 4p^4(1-p) + 10p^4(1-p)^2 + 20p^3(1-p)^3\, p^2/\{p^2 + (1-p)^2\}$$

If the two players are evenly matched then $p = 0.5 + u$ where u is very small (much less than ½, so we can ignore powers like u^2 compared to u), and we find that approximately $G = ½ + 5u/2$.[2] If the players are equally matched then u is zero and $G = ½$ so they have an equal chance of winning the game. However, if u is a little bigger than zero then we see that the marginal advantage, u, that one player has in winning each point grows to a much bigger chance, $5u/2$, of winning the whole game.

We could just carry on to work out the chance of winning a second game, and then to win a set, S.[3] Overall, a small advantage u in winning a point translates into an advantage of about 11u in a three-set match and 13u in a five-set match. Small differences in ability soon make long matches poor contests despite the scoring system's attempt to reinvent the match by the sequence of games and sets.

Let's change our approach to include psychological factors. If you win the first set then this boosts your confidence and allows you to play more adventurously than if you have lost the first set and are facing overall defeat if you lose the second. If you play a best-of-five match and win the first two sets then you have an even greater psychological advantage over your opponent which should make it easier to win the third set and clinch the match. The odds, O, on you winning the first set are the probability that you win a set, S, divided by the probability that you lose it, so $O = S/(1 - S)$. But let's suppose that if you win a set you get a psychological boost B and your odds of winning the next set are now $O \times B$, but if you lose those odds fall to O/B. Then, if you win the second set, your odds of winning the third grow by another psychological B factor to $O \times B^2$. The tree of possibilities for you in a best-of-three set match is shown here:[4]

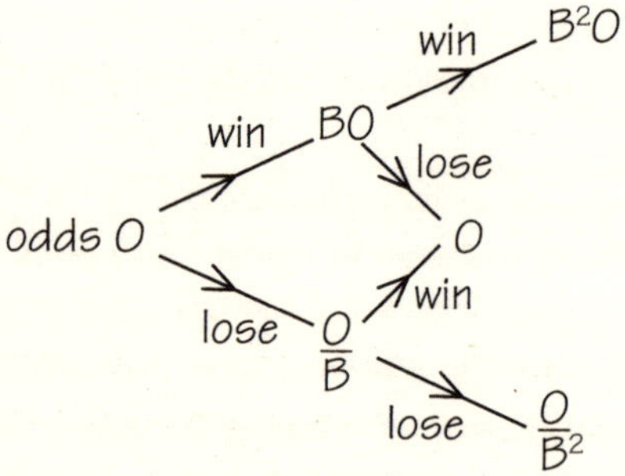

When B = 1 there is no psychological advantage to winning a past set but when B is larger than 1, the gain grows strongly with each win. The odds on you winning by 2–0 is B^2 times the odds before the start of play, and your odds of winning 2–1 are B times the starting odds. This simple model is quite different to the one we used to work out the way points advantage transfers to game advantage. In reality, some elements of both, together with a small random element, and also the inclusion of the big increase in the probability of winning a point on your serve (more than 2 in 3 for top players), will blend together to create a more complete picture of how physical and psychological factors impact on the result of a progressive game like tennis.

80

Goals, Goals, Goals

There are four Olympic sports that have goals. Each has the aim (some might say the goal) of getting a ball to completely cross a bounded rectangular area known as the 'goal' while a goalkeeper tries to frustrate these efforts by catching the ball or deflecting it wide of the goalposts. The four goal-directed sports of football, water polo, handball and field hockey are played on and above pitches of very different sizes and composition; the balls too differ widely. They all share the concept of a penalty shot where a single attacker occasionally gets to take an unhindered shot at the opponent's goal with only their goalkeeper to defend it.

In the table below I have provided various measurements for these four sports so that you can compare the degree of difficulty involved in scoring a goal in each of them. The first three columns give the height, width and area of the goal that the attackers have to shoot at. I have specified all these dimensions in metres and you will notice that the modern game of handball that was created in continental Europe has its goal dimensions neatly defined in metric units. The other three sports have English origins and their goal dimensions were originally defined in Imperial units of feet or yards (for example, in football the goal is 8 yards wide and 8 feet high) and so look peculiar when converted to metric units. In the fourth column I have listed the diameter of the playing ball in centimetres. This is what the attackers have to shoot through the frame of the goal. In the sixth column I have given the distance from the penalty spot to the goal. The seventh column provides the cross-sectional area of the ball (X–S) and the next one the ratio of the area of the goal to the area of the ball. This indicates how much room there is available for a scoring shot. Finally, I have given the ratio of the distance to the penalty spot divided by the square root of the goal area. This last quantity, $P/\sqrt{A}$, the 'Penalty Factor', is a number without units that measures how hard it is to score from a penalty shot. Larger values of $P/\sqrt{A}$ mean it is harder to score. This is either because P is getting larger and the penalty spot is further away, or because A is smaller and the goal target area is reduced. Conversely, the smaller $P/\sqrt{A}$, the easier it is to score. Interestingly, despite the huge variation in goal and ball sizes, and the range of distances to the penalty spots in these four very different sports, the Penalty Factors are very similar. The ladder of difficulty for scoring a penalty, starting with the easiest and rising to the hardest, is therefore field hockey, football, handball and water polo. But it is the close similarity that is most striking.

Sport	Goal height, m	Goal width, m	Goal area, A/m^2	Ball diameter, cm	Penalty distance, P/m	Ball X–S area, B/m^2	Ratio A/B	$P/\sqrt{A}$ Penalty Factor
Football	2.44	7.32	17.86	22.0	11	0.038	470	2.6
W. Polo	0.9	3	2.7	22.0	5	0.038	71	3.0
Hockey	2.14	3.66	7.83	6.8	6.4	0.0036	2175	2.3
Handball	2	3	6	18.8	7	0.028	214	2.9

81

Total Immersion

Freestyle swimmers must overcome many forms of resistance offered by the water in their quest to move as swiftly as possible down the length of the pool. Unlike runners or cyclists, swimmers don't propel themselves by pushing off solid objects, like pedals or the ground, except momentarily at the start and the turns. More than 85% of the front-crawl swimmer's thrust comes from the work done by the arms and hands, and the speeds achieved are roughly four times smaller than those for runners over the same distance. The hand acts like a hydrofoil in the water, generating lift but also creating drag, and both of these contrary forces are proportional to the density of the water, the swimmer's speed and the surface area of the hand.[1] The lift pushes upwards and the drag opposes the forward motion. Their sizes depend delicately upon the angle at which the hand cuts the water and then sweeps backwards through it. There is much room for coaching input and careful optimisation of the swimming stroke.

There are three main sources of drag resisting the forward motion of the swimmer. They are created by friction, pressure and water waves. The frictional drag arises in a thin layer of water very close to the swimmer's body. It is largest if the flow in that layer becomes turbulent, rather than smooth, and this is determined by the speed and size of the swimmer and also how streamlined and smooth his body is in the water.[2] As we have already seen in other cases of movement through resisting media, this drag force is also proportional to the square of the swimmer's speed and his cross-sectional area in the direction of motion.

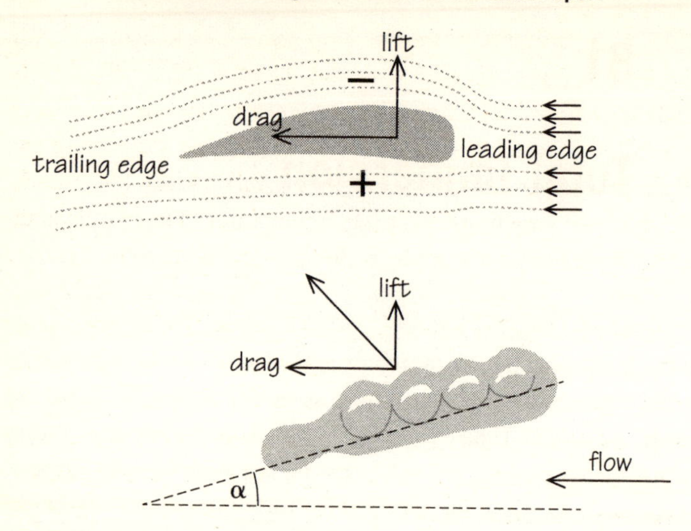

The second type of drag is created by pressure and arises because fast swimmers create a pressure difference between the water immediately in front of them (higher pressure) and behind them (lower pressure). The strength of the drag depends on the pressure difference times the cross-sectional area of the swimmer and the square of the swimming speed.

The third main type of drag, due to waves, affects swimming near the water's surface. Some of the swimmer's energy goes into making waves. As the swimmer speeds up, and generates more energy, so the amplitude and length of the waves from peak to peak increases.[3] When these waves have a length a little greater than the height of the swimmer, he will find himself caught in a wave trough of his own making which stops him generating more speed. The smaller the swimmer the lower the speed at which this problem is encountered. However, by refining the stroke technique a very good swimmer can reduce the amplitude of these surface waves. You can also remove this form of drag completely for part of the race by swimming a significant portion of each length underwater using a dolphin leg kick.

This is why freestylers take a long time to surface after the start or a tumble turn from the wall.[4] In 2010, the American swimmer Hill Taylor broke the world 50m 'backstroke' record by a full second for the crowd's amusement by swimming the whole distance underwater using a dolphin kick and his remarkable ability to streamline his body underwater.[5] He was disqualified so his 23.1s didn't replace the previous backstroke record of 24.04s.

These three forces come into play one by one as the speed of the swimmer increases. At very low recreational swimming speeds, the frictional drag is the most important. Your speed will be too low to create a significant pressure difference over the length of your body and no long waves will be made either. As you start to swim faster you will begin to create a pressure reduction behind you and an increase in front of you and the pressure drag produced by the difference will eventually become larger than the frictional drag. As you increase speed further, and approach the 1.5–2m/s speeds of top swimmers, the long waves you generate will make wave drag more and more important. The total drag force at speeds, V, exceeding 1.3m/s will be about $40(V/1.3m/s)^2N$, and the wave drag will be about 0.56 times this.[6]

A quick look at the world record progression in swimming shows how big an impact this type of scientific understanding of swimming has had on the efficiency of the stroke process that was led by coaches like the American James Counsilman in the 1980s. In 1964, the world records for 100m freestyle were 52.9s for men and 58.9s for women. Today they are 46.91s and 52.09s respectively, giving advances of 6s for both sexes. The top women are swimming faster than the men's Olympic champion of 1964, and a superstar of the sport like Mark Spitz, who swam a world record of 51.22s when winning one of his eight gold medals at the Munich Games of 1972, wouldn't even qualify for the US Olympic trials today.

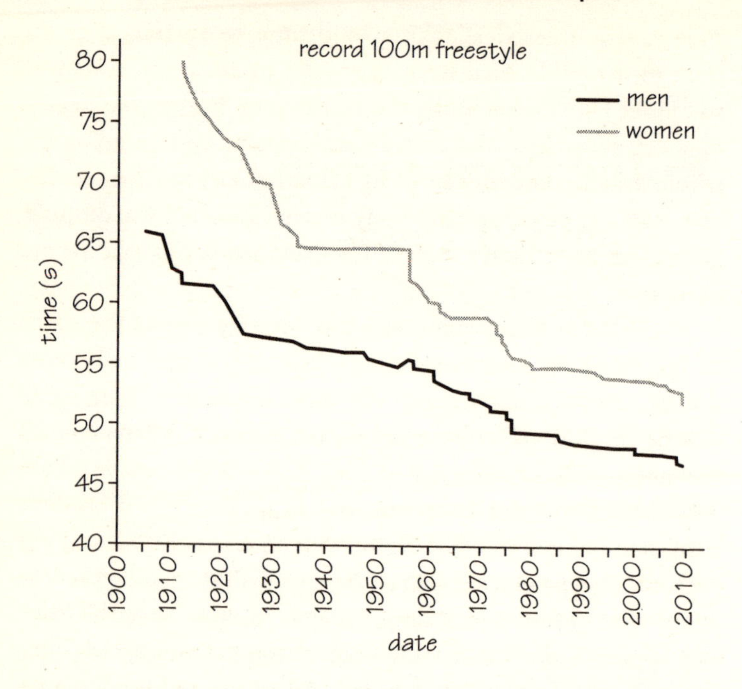

In track athletics the 400m run lasts for a comparable time and provides an interesting comparison. The 1964 world records were 44.9s and 51s for men and women respectively. Today, those record times have advanced to 43.18s and 47.6s. These advances are 1.7s and 3.4s respectively – way below the 6s progress in the swimming records over the same period despite the advent of all-weather running tracks. This is a measure of the huge possibilities that were available for optimising the swimming stroke compared with the more limited advances in running speed made possible by improved technique.

82

The Great British Football Team

One of the great mysteries of the universe is why Britain doesn't want to play in the Olympic football tournament. Once upon a time it did. There was no football competition in the 1896 Games but Britain was represented at the 1900 Paris Olympics by Upton Park Football Club – no connection with West Ham United or their present ground of the same name[1] – which won the tournament, beating France 4–0. Only Britain, France, Belgium, Germany and Switzerland had entered teams and the last two countries later withdrew. In the 1904 Olympics a football club from St Louis and a Canadian team were the sole participants: it was simply too expensive for teams to travel all the way to St Louis from Europe.

Britain was persuaded to organise a football competition for the 1908 Games in London and the lower travel costs attracted eight entrants, all from Europe, with two teams from France. Hungary and Bohemia then withdrew and this gave the Netherlands and France 'A' byes into the next round. Britain won the gold medal, beating Denmark 2–0 in the final in front of about 8,000 spectators at the old White City Stadium in Shepherd's Bush. The results of all the games played were as follows:

Quarter finals

Netherlands (bye) beat Hungary (withdrew)

France 'A' (bye) beat Bohemia (withdrew)

Great Britain 12	Sweden 1
Denmark 9	France 'B' 0

Semi finals

Great Britain 4	Netherlands 0
Denmark 17	France 'A' 1

Final

Great Britain 2	Denmark 0

Bizarrely, in the third-place match the Netherlands beat Sweden 2–0 because France 'A' had withdrawn.

Incidentally, the silver-medal-winning Danish team included one Harald Bohr, a pure mathematician and the brother of the great physicist Niels Bohr, who was himself an accomplished goalkeeper. Harald scored twice in Denmark's opening game. At the 1912 Stockholm Olympiad, eleven teams participated and Britain again beat Denmark 4–2 in the final.

Britain entered the 1920 tournament but was eliminated 3–1 by Norway in the first round of a knockout competition. The whole event ended very badly. For the only time in football history the final, between Belgium and Czechoslovakia, of a major competition could not be completed. The Czechs walked off the field near the end after one of their players was sent off, complaining about bias by the officials (who were all English) and intimidation by Belgian soldiers in the crowd; the game was awarded to the Belgian hosts by default who were leading 2–0 at the time.

Britain didn't compete[2] in the football tournament again until the 1936 Olympiad and then again on home soil in the post-war London Olympiad of 1948, where the British team was managed by Matt Busby and came fourth. In 1952, 1956 and 1960 Britain lost in the early rounds, and over the period 1964–72 they didn't even manage to qualify for the competition. They never entered the Olympic football competition again.

Why not? It seems strange to outsiders that Great Britain, the 'home of football', doesn't even enter the Olympic tournament. This looked especially odd when the city of Manchester was put forward as the British candidate to host the Olympic Games of 1996 or 2000 with its bid being fronted by famous footballers like Bobby Charlton and David Beckham. This anomaly seemed to pass everyone by at the time and erudite deliberations about the politics of the decision not to award the Games to the British city seemed to miss this crucial point entirely.

The answer to the puzzle of Britain's footballing absence is political. There are three threads to it. For historical reasons, the four home countries – England, Northern Ireland, Scotland and Wales – are separate FIFA members. As a result, FIFA has more members than the United Nations! The eight-man International Football Board that controls the rules of the game has four FIFA members and also one from each of the four home countries.

Therefore, when the Olympics first allowed professionals to compete with amateurs after 1974, Britain withdrew from the football competition. It feared that fielding a British team in this tournament would lead to pressure from other FIFA members to merge the four home nations into a single professional British team. The players didn't like the idea because they would suddenly be four times less likely to take part in an international competition. Nowadays, the single team would consist almost entirely of English players. In addition, old rivalries make the merger anathema to many players, supporters and nationalistic political parties, like the Scottish National Party and First Minister Alex Salmond, who oppose the idea on much wider political grounds. Another more subtle impediment is the realisation by political allies of the home nations within FIFA that four votes would suddenly become one. In recent shenanigans about corruption in the higher echelons of FIFA, an English resolution to delay the vote on Sepp Blatter's re-election in 2011 found support from Scotland but not from Wales and Ireland who were perhaps fearful that their separate status would be challenged if they rocked the boat.

Still, everyone expected that even these fears would be put to one side for the 2012 tournament. As hosts, Britain did not have to qualify for the tournament and a unified team might appeal to the public. However, despite encouragement by the prime minister of the day and the chair of the British Olympic Committee, the Scottish, Welsh and Irish football associations rejected the idea of joining a British team. They didn't believe FIFA's assurances that no precedent would be created that would lead them to be subsumed by the English Football Association. However, it is still open for individual players from these countries to accept an invitation to play, although their clubs are unhappy about them playing these extra games. The only alternative seems to be for an all-English team to play under the name of Great Britain. We shall see!

83

Strange But True

You manage a sports club with teams at many levels – first, second, third, juniors, all the way down to the beginners. Your club is huge and there are many layers of management structure. There are always decisions to be made about promoting people, lower-team players and middle managers, all of whom are aspiring to move up the ladder. All large organisations are the same. Given two candidates, one very successful in their current job but the other less competent, most people would regard it as a 'no-brainer' that you promote the first one in preference to the second. However, the Canadian psychologist Laurence Peter introduced his famous 'Peter Principle' in 1969 as an argument against this commonsense approach: people will not necessarily work or play as well at the next level up the ladder where different skills may be required.[1]

One obvious result of the commonsense promotion strategy is that everyone will move upwards until they find themselves in a job they are no good at. They won't be promoted again and the whole organisation will have incompetents in every position! Thus Peter argued that every new member of an organisation climbs upwards until they reach their level of maximum incompetence.

At root, Peter's Principle is based on the assumption that there is little or no correlation between competences at different levels in an organisation. The commonsense view assumes that there is but its validity is not so clear. Good teachers don't necessarily become good head teachers, good footballers don't necessarily become good managers, good academic researchers may not be good university vice chancellors, good athletes may not be good coaches, and good medical students may not become good doctors.

Recently, there have been studies using computer models of organisations to test the results of different promotion policies.[2] Strikingly, they found that the policy of always promoting the least competent person can maximise the overall competence of the organisation – and promoting the most competent candidate can significantly reduce it.

	Promote the best	Promote the worst	Promote best and worst at random
There is correlation between different job requirements ('commonsense view')	+9%	–5%	+2%
There is no correlation between different job requirements ('Peter Principle')	–10%	+12%	+1%

The table shows the results of experiments by Alessandro Pluchino, Andrea Rapisarda and Cesare Garofalo at the University of Catania. They calculated the change in overall organisational effectiveness of the three policies – 'promote the best', 'promote the worst', or 'promote at random' – to examine the correlation between the abilities needed to perform in your existing job and in the one you are promoted to do. We see that if that correlation doesn't exist, as Peter assumed, then 'promote the best' leads to a 10% fall in effectiveness across the organisation, while 'promote the worst' leads to a 12% gain. Even if there is a correlation, 'promote the best' only leads to a 9% gain. In the last column are the results of choosing at random between the best and the worst candidates. There are of course other possibilities like promoting the best for half of the positions and the worst for the others, or some other mix of the best, worst and random strategies.

All this is very worrying for football managers, sports administrators and chief executives, and totally contrary to their intuitions. And of course, human actors can behave unpredictably when new policies are introduced. If the worst candidates get promoted there will be a race to the bottom to convince your boss that you are the most incompetent and therefore the most deserving of promotion!

84

Blade Runner

One of the most unusual disputes in sport has sprung up around the remarkable South African sprinter, Oscar Pistorius. Oscar's running times are 45.07 for 400m, 21.41s for 200m, and 10.91s for 100m – yet he is a double amputee below the knee since he was eleven months old. He has also been successful in rugby, water polo, tennis and wrestling but he has specialised in track sprinting, winning gold medals in the 100m, 200m and 400m at the 2004 Paralympics, following up with a 100m bronze and 200m gold again in 2008. One of his aims has been to gain the Olympic qualifying time in his best event, the 400m, and compete against able-bodied runners in the London Olympics.

But do his artificial blades give him an unfair advantage over able-bodied athletes? The athletics governing body, the IAAF, commissioned a comparative study of Pistorius' 400m running mechanics compared with five able-bodied runners of similar height and bodyweight who were able to run at comparable speeds (46.5–49.26s) over 400m.[1] Not surprisingly, the researchers found that the mechanics of the artificial blades was rather different to human limbs. In the able-bodied athletes only about 40–45% of the energy stored in their ankle joints was returned to aid the next phase of the running movement. In Pistorius' case the elastic blade returned 90–95% of the stored energy and so the knee joint contributes very little, at most 5%. In addition, blades don't tire. As a result of this experimental analysis he was judged to possess an unfair advantage and was not permitted to compete in the able-bodied Olympic 400m in 2008. However, as he didn't qualify for the South African team in the end, the IAAF's decision didn't have any effect.

Pistorius' legal team then appealed the IAAF's decision to the Court for Arbitration in Sport (CAS) in Lausanne, who overturned the ban and cleared the way for him to compete at future World Championships and the 2012 Olympics. This decision was rather unsatisfactory as it merely addressed one technical point on which the IAAF judgement had been based – the claim that Pistorius' blades gave him an ability to maintain speed over a 400m race. The CAS judged that the IAAF evidence did not support the ban.

Eighteen months later, a major argument broke out. The scientific team which had provided the evidence to support Pistorius' appeal, led by Peter Weyand at Rice University in Texas, published their studies of his running.[2] Although they found no evidence for the specific advantage claimed by the IAAF, they found strong evidence for another major advantage created by the differences between able-bodied and amputee running mechanics. The top speeds attained by the able-bodied athletes were similar to Pistorius', but the time it took to reposition the limb for the next stride was much less (0.1s) for the amputee than for able-bodied athletes, aided by the elasticity and the fact that the mass below the knee was half that of an able-bodied runner. In fact he was 18% quicker in this regard than the last five 100m world record holders. Weyand and his colleagues therefore concluded that 'Pistorius' sprinting mechanics are anomalous, advantageous and directly attributable to how much lighter and springier his artificial limbs are. The blades enhance sprint running speeds by 15–30%.'

These conclusions have created something of a furore, especially as Pistorius has said that he would retire from able-bodied competition if he thought he was gaining an unfair advantage and he competed again in the 2011 World Championships. Some have argued that they should have revealed those facts to the CAS in 2008, yet Weyand says they were only free to comment on what was in the public record. There have also been a series of objections to the Weyand research that shows that this issue is far from settled.[3]

Yet there are two observations that might be made about all this from running experience rather than laboratory studies. The claim that Pistorius' 400m running is aided by 15–30% by his blades is not credible; it would mean that if he was able-bodied he would be running between 53.6s and 65.1s! The idea that someone with his general physical attributes might not be able to run faster than 65s for 400m if able-bodied is silly.

The second point is something that has been overlooked by all the researchers and seems to weigh strongly against the conclusion that Pistorius' ability to maintain his speed over longer distances is identical to that of limb-intact athletes. The key evidence against this conclusion is the relation between Pistorius' 200m and 400m times. No able-bodied runner who has run 45.1s for 400m would only be able to run 200m in 21.4s. If we just take a dozen British athletes on the all-time performance list for 400m who have run between 45.63s and 45.85s, then we find the range of 200m performances is 20.76–21.01s. Pistorius' 400m times are far quicker than those able-bodied runners with the same 200m performances can achieve. This suggests, contrary to the CAS ruling, that his blades *do* significantly help him maintain a fast tempo over the second half of the 400m race – although there are different ways in which he might be doing this. His 100m and 200m times reveal that he is clearly hindered in getting up to speed after the start. Unlike top able-bodied athletes, however, who almost all run the second half of their 400m races between 0.7–2.7s slower than the first, Pistorius often runs the second half more than a second faster. Ideally, we need performance data for Pistorius over 500m and 600m distances. The advantage provided by his blades should become far more evident and he might even challenge the world records for these distances. It is strange that all the biomechanical testing didn't think to carry out this crucial test over longer distances.

85

Pairing People Up

Imagine you are a coach or a competition organiser and you have to ensure that all your athletes play against each other in a series of sessions. When the numbers are small this is pretty easy to set up but as the number of players grows it becomes trickier. Is there a system you should use to make sure that at each successive session everyone has a partner they haven't played before?

Let's be specific and assume there are fourteen players to be organised into sequences of seven matches and they must all play each other once only. We will label the fourteen players by letters: A, B, C, D, E, F, G, H, I, J, K, L, M, and N.

Write each letter on a square label and lay them down in two adjacent rows of seven, like this:

A	B	C	D	E	F	G
H	I	J	K	L	M	N

The pairings for the first session are the vertical columns: A vs H, B vs I, C vs J, and so on. To get the next session pairing, *remove* the label marked H, drop A down into its place, slide all the remaining letters in the top row one place to the left to fill the gap left by A, move N up into the top row slot just occupied by G, slide the remaining letters on the bottom row to the right, and put H in the empty space. The new pattern looks like this:

B	C	D	E	F	G	N
A	H	I	J	K	L	M

The second session of match pairings are the new vertical couples: B vs A, C vs H, D vs I, and so on. To get the next round just repeat the process, first removing A, dropping B down and shifting everyone anticlockwise before dropping A back in the empty space left between B and H. Keep on doing this until everyone has played everyone else. The system works for any even number of players. It works for dancing partners too!

86

Ticket Touts

Two ticket touts are selling tickets to a football match. They each have 30 tickets and Del is offering 2 tickets for £100, while his rival, Rodney, is offering 3 for £200. Worried that they might be forced to cut their prices in a competitive race for market share as kick-off approaches, they decide to join forces, put their tickets together in a pool of 60 and offer them at 5 for £300. What have they gained by this corporate merger? If they had sold them all before collaborating they could have had a total income of £3,500, or £1,750 each, but their total income after they merge will be £3,600, or £1,800 each.

A rival pair of competing touts, Sharon and Tracey, who are offering tickets at 2 for £100 and 3 for £100 respectively, and also have 30 each, hear about Del and Rodney's winning strategy and decide to do the same, adding the ticket numbers and the prices. They offer their 60 tickets at a price of 5 for £200. They confidently expect they are also going to do better. But soon they do a little arithmetic. Their maximum total income before they merged was £1,500 + £1,000 = £2,500 but after they joined forces this maximum falls to £2,400, or £1,200 each.

There is a general rule here that we can pick out. If the bigger price per ticket is A tickets for £B and the smaller is C tickets for £D (so $B/A > D/C$), where we had $A = 3$, $B = 200$, $C = 2$, $D = 100$ for Del and Rodney, then joining forces and selling at a price of $(B + D)/(A + C)$ per ticket does better than the average price when selling separately, which is $\frac{1}{2}(B/A + D/C)$, so long as

$(B/A - D/C) \times (A - C)$ is positive. This requires the prices originally offered per ticket to be different (B/A must not equal D/C) and also that A is bigger than C. We see that Del and Rodney's cartel meets this requirement because $A = 3$ and $C = 2$ but Sharon and Tracey have $A = 2$, $B = D = 100$, and $C = 3$ so $A < C$ and they fail to meet the condition for profitability.

87

Skydiving

Skydiving is leaping out of high-flying aircraft, then falling freely through the air at high speed until you open your parachute and descend gracefully to the ground. In practice, it's a bit more demanding than I have made it sound, but some simple mathematics reveals the essentials. When you jump out of the aircraft, or from a balloon gondola, you feel two forces: your weight acting vertically downwards through your centre of mass, and the drag force of the air which acts upwards. Your weight is equal to mg, where m is the mass of your body and your kit, and g is the acceleration due to gravity. The opposing air drag is equal to $D = \frac{1}{2}C\rho Av^2$, where v is your speed of descent, ρ is the density of air (whose variation with altitude we ignore here), C is the drag factor for motion through air, and A is the projected area of your body shape perpendicular to your downward direction of motion.

What happens after you leap from the plane is that your weight will accelerate you and cause your speed of descent to increase quickly. The increasing speed of descent causes the drag force to increase more quickly because it is proportional to the square of your speed relative to the air. Soon the two forces will become equal in magnitude, but they act in opposite directions so the net force on the skydiver will become zero. He is then acted upon by no net force and will cease to accelerate or decelerate: he will just fall at a constant speed. This is called the 'terminal velocity' – so called because it is the end result of falling through a resisting medium rather than because of any premonitions about parachutes failing to open. If we equate the weight mg to the drag force D we find the value of v which defines this terminal velocity to be[1] $U = (2mg/C\rho A)^{1/2}$.

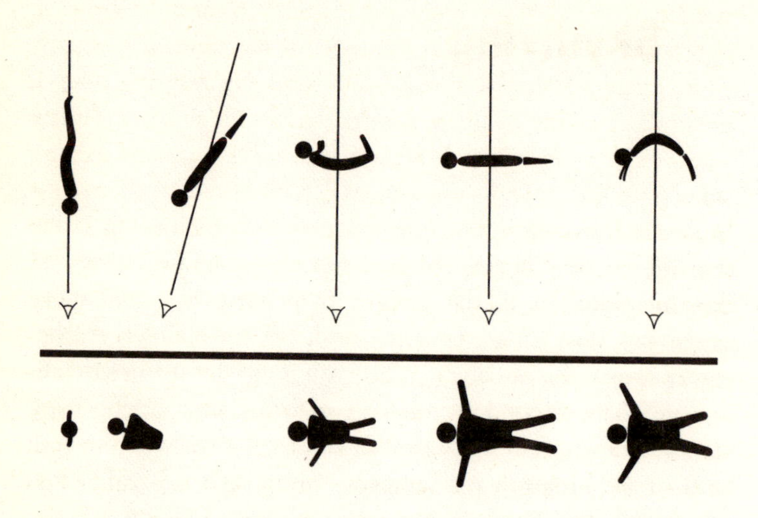

This formula shows us some interesting things. The terminal speed U gets larger as your projected area, A, gets smaller. When you watch a group of expert skydivers falling in formation they will start at different altitudes, then catch one another up and then maintain the same altitude whilst joining hands or performing other manoeuvres. Different body positions create different projected areas which are shown in the figure above. If you change your body position to become more upright you can reduce the value of A, fall with a higher terminal velocity and catch other skydivers who have spreadeagled their bodies to achieve a larger resisting area and a lower terminal speed. A skydiver with a weight of 60kg will reach a terminal speed of about 50m/s in a spread position but he could reach in excess of 80m/s in a streamlined head-first configuration. We can also see that the terminal speed depends on the weight of the skydiver. The heavier you are, the faster you will be falling when you reach your terminal velocity.

The next and most vital stage of the skydiving experience is to open your parachute. This increases the air drag dramatically by suddenly increasing A to about $25m^2$. Previously the 600N weight of the skydiver was exactly counterbalanced by the air drag when falling at speed U, but now the large area of the parachute canopy falling at speed U increases the resisting drag. It greatly exceeds the weight and so decelerates the skydiver towards a new terminal speed that is around 10m/s. Old wartime pictures show parachutists with large curved parachutes that look circular in profile from below. They have a hole in the centre to let air through in order to counter the tendency to flip over as the air tries to avoid the resistance created by the fabric by rushing sideways. Nowadays parachutes are square. This considerably reduces the area of fabric surface needed to provide the required cross-sectional area with the air. It also makes the parachute far more stable and manoeuvrable because there can be cords tied to each of the four corners of the parachute which skydivers can pull on independently to fine-tune their descent and landing.

Finally, a look in the record books reveals that almost all the records for this activity are held by the legendary American air force pilot, Joseph Kittinger.* In 1960, he skydived from a helium balloon at 31,330m, fell freely for 4m 36s and reached a speed of 255m/s, experiencing temperatures of –70°C. He opened his parachute at an altitude of 4,270m and descended to the ground. He set records for highest balloon ascent, highest parachute jump,[2] longest free-fall descent and the highest speed achieved by an unaided human moving through the earth's atmosphere. And he holds another record. During one of his super-descents equipment failure put him in a horizontal spin that reached 120rpm. This meant he experienced a centrifugal force[3] of more than 20g, the highest recorded. Even Superman would struggle to compete with these achievements.

*Felix Baumgartner set a new free-fall altitude record of 39km on 14 October 2012

88

Running High

The choice of Mexico City to host the 1968 Olympic Games first brought the word 'altitude' into the vocabulary of athletics. Mexico City sits at an altitude of 2,240m above sea level. This caused numerous problems. For the distance running events above 800m, it was much harder to run at altitude because oxygen absorption by the body was diminished by 10–15% for unacclimatised runners. Athletes who lived at altitude, notably Africans, were significantly advantaged and won all the distance running events. However, their winning times were generally poor compared with their own achievements at sea level and there was a general feeling that many of the world's best endurance athletes had been unfairly prevented from winning or setting records by the effects of altitude.

For the sprints and horizontal jumps, on the other hand, the lower air density meant less air resistance and faster times. World records were set in almost all the men's and women's sprints, horizontal jumps and relays in Mexico City. Here the psychological problem for athletes was that they might never approach those performances again at sea level. As a result of the controversy that sprang up about unequal competition and altitude-aided records at the Mexico Games, performances at altitude began to be distinguished from those achieved at sea level.*

*The 1968 Games were the first to use fully automatic electrical timing to 0.01s accuracy and thereafter 'hand-timed' performances (which always seemed to be faster) were distinguished from electrical timings in performance tables. The Mexico Games were also the first in which drug testing took place, and in which the athletics events were held on an artificial all-weather track surface, rather

Why does altitude help the sprinters? The drag on a runner moving at speed V through air with a following wind of speed W of density ρ is proportional to $\rho(V - W)^2$. We can see immediately that, all other things being equal, a decrease in the density of air will reduce the drag and lead to more of the runner's power being used for fast forward motion. At sea level, the air density is 1.23kg/m^3, but at the 2,240m altitude of Mexico City it is down to 0.98kg/m^3, at a moderate temperature of about 20°C. This means that the drag on runners in Mexico City because of air resistance was smaller by a factor 0.98/1.23 = 0.8 than at sea level. This should lead to a time improvement of about 0.08% in events like 100m, 200m and 400m. This is significant but it is not large enough to explain the 1.5–2% improvements displayed by male and female athletes in those events at the Mexico Olympic Games.[1]

The answer to this discrepancy could lie with the wind. The wind readings in the jumps and sprints in Mexico City were unusually favourable. A suspiciously large number of world records were set with the wind officially recorded at or near +2m/s – the maximum allowed for record purposes. The wind was at the 2m/s level for world records in the women's 200m and three world record jumps (on two different days) in the men's triple jump, as well as for Bob Beamon's famous long-jump record. It was at 1.8m/s for the women's 100m record. A 2m/s following wind will improve a 100m time by about 0.11s compared to still air. Its effect is proportional to the square of the speed and is to reduce the drag factor from V^2 in still air to $(V - W)^2$ when there is a tailwind of W. The high altitude was only cutting times by about 0.03s per 100m, so the wind could be much more important than the altitude.[2] In combination they provide some understanding of the improvements in the 100m, 200m and horizontal jumps.

In retrospect, it was the impact of altitude on the distance

---

than on cinder tracks. These tracks produced consistently faster times in all events. Overall, the Mexico Games were revolutionary for track and field athletics in many ways.

events that left the greatest legacy. There would never be another high-altitude Olympics because of the problems experienced by athletes who lived and trained at sea level and the vocal criticism of the evident unfairness.* But sea-level athletes recognised the benefits that could be gained by spending extended periods living and training at high altitude. Ever since the 1968 Olympics, leading distance runners have tried to gain some of the physiological benefits bestowed upon athletes born at high altitude, notably Kenyan runners. Training at high altitude conditions the body to absorb oxygen at lower pressures. When an athlete then returns to sea level the body will absorb a lot more oxygen into the bloodstream than is usual for them at sea level and their performances will be boosted. The effect is only temporary and careful timing of the period of altitude training and the time when you return to sea level before competing is needed. Some athletes have taken these things to extremes by sleeping each night in oxygen tents which simulate the atmosphere at high altitude. There are also still occasional rumours about athletes undergoing 'blood doping' to bottle the benefits of high-altitude training. Some of their blood would be removed at a time when the oxygen-carrying capacity had been maximised and then transfused back into the body later on before a big competition. While theoretically beneficial, and invisible to drug tests, one suspects that most athletes would be unwilling to undergo blood transfusions as part of their last-minute preparation for the Olympic Games.

*World records cannot be set at altitudes exceeding 1000m.

89

The Archer's Paradox

There is something paradoxical about shooting an arrow at a target. You can't hit the target by aiming straight at it: the bow shaft gets in the way. The arrow cannot be aligned through the middle of the bow, so it must be on one side of the shaft or the other (depending on the handedness of the archer). After the arrow is released by the fingers, the bowstring pushing on the nock of the arrow does not propel a perfectly flighted arrow straight towards the target: a correctly aimed arrow shot by a right-handed archer should end up well to the left of the target.

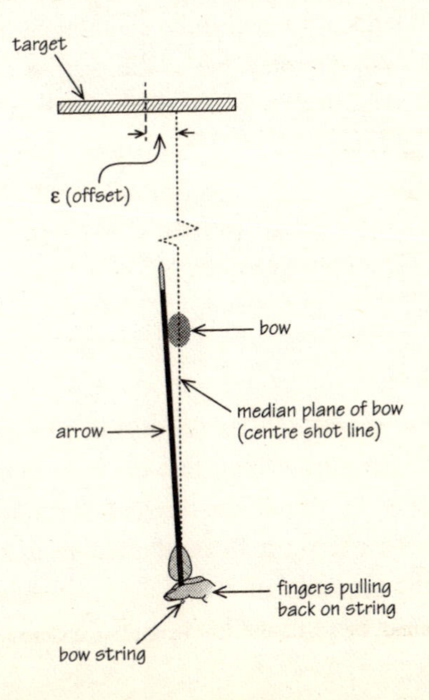

This is the 'Archer's Paradox'. It was well appreciated from experience by archers for hundreds of years before the advent of modern high-speed photography enabled very close observation of what happens to the arrow.

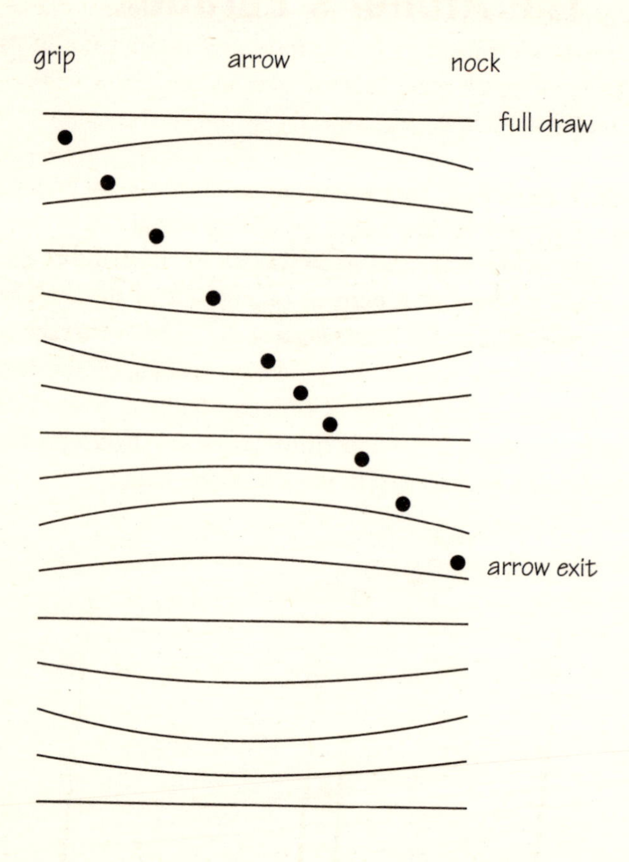

When the arrow is released by the archer's fingers it receives a sudden impulse that pushes it against the bow shaft. This contact, and also the way in which the fingers are released from the bowstring, causes the arrow to flex and undulate as it moves forward at high speed (the sequence of arrow shapes is shown in the figure above). The archer can't move his fingers fast enough to stop the effect of the finger release, and the glancing blow from the bow shaft is totally unavoidable. The frequency with which the arrow undulates depends on how its mass is distributed along its length and how stiff it is.[1] This stiffness is called the 'spine' of the arrow and getting it right is crucial. Too little spine (more stiffness) and the arrow won't flex and will be forced off to the left as it passes the bow; too much spine (less stiffness) and the arrow will deform so much that its flight will be deflected off to the right.

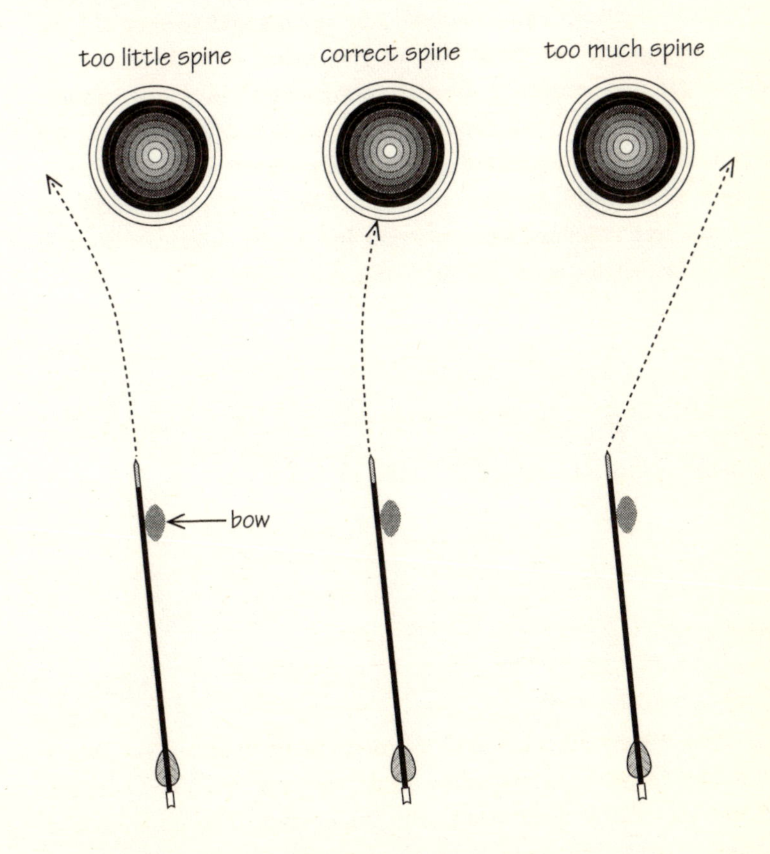

In between, there is a 'Goldilocks' arrow spine that gets it just right, and the resulting undulation in the shape of the arrow exactly counters the deflection created by the bow shaft and the arrow after it snakes around it. When this undulation of the arrow is completed it flies straight towards the intended target. The arrow actually continues oscillating en route to the target but the strength of the undulations is strongly damped down after it flies freely and they soon have no significant effect. The parabolic fletches[2] at the end of the arrow serve to counter the arrow's tendency to keep turning in one direction in flight by contributing air drag against sideways and rotational movements and so they aid the straightening. The great skill of the archer is to determine by practice and sensitivity the right combination of arrow stiffness and weight, bow set-up and finger release that will ensure that the undulation of the released arrow exactly counters the initial deviation. Harder still, in competition you have to be totally consistent in the way you draw the bowstring and release the arrow on each shot so that the same tuning works every time.

90

Bend It Like Beckham

One of the skills that soccer players seek to master from an early age is how to 'bend' the ball. For those ignorant of the game, we should say this simply means learning how to kick the ball so that it swerves in the air. This skill can deceive opposing defenders and goalkeepers and is especially potent when a free kick is awarded close to the edge of the opponents' penalty area. The defending team will set up a wall of players at the requisite ten yards from the ball (closer if they can get away with it) to block a direct shot at their goal. The attacking team may well have a prized player, like David Beckham, who can bend the ball around or over this defensive wall so that the initial trajectory of the ball heads past the edge, or over the top of the wall of defenders, only for it to swing back in, or down, and end up in the goal. Alas, the poor goalkeeper finds that his wall of defenders does nothing more than block his view of the ball until it is too late to respond. How is this done and why is it possible?

When you kick a football off-centre it will spin. If you kick the right-hand side of the ball with the inside of your right foot then it will spin in an anticlockwise direction, but kick it on the left-hand side with the outside of your right foot and it will spin clockwise. The greater the spin you can put on the ball, the more it will swerve.

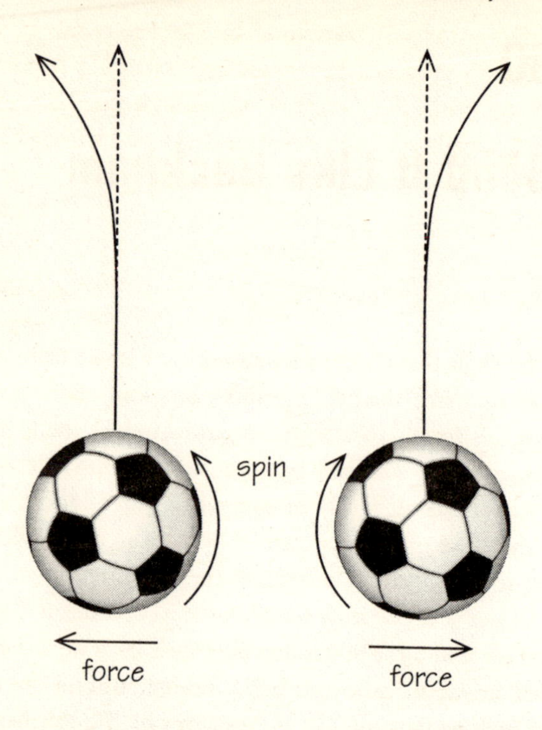

The most famous example of this skill was a free kick taken by Roberto Carlos for Brazil in a 1997 match against France. He struck the ball initially from a spot 35m from goal, close to the corner of the D on the edge of the penalty area, at a velocity of about 130kph. The subsequent swerve was so dramatic that a ball boy standing 10m to the side of the goal jumped out of the way because he thought the ball was heading straight for him before it swerved suddenly away from him and beat the goalkeeper. Carlos hit the ball so hard that gravity never had a chance to damp down the aerodynamic motion.

This effect is not unique to soccer. It can be seen across the whole sporting spectrum. Volleyballs, baseballs, cricket balls, tennis balls – all can be given a spin that will result in them following a curved trajectory which a non-spinning ball would not. The reason for the swerve can be understood by looking at the air flowing past the ball. In the figure below, the ball is moving to the left and not spinning. When the airflow impinges on the ball the flow lines are pushed together, so the pressure drops and the speed of the air passing the surface of the ball increases.

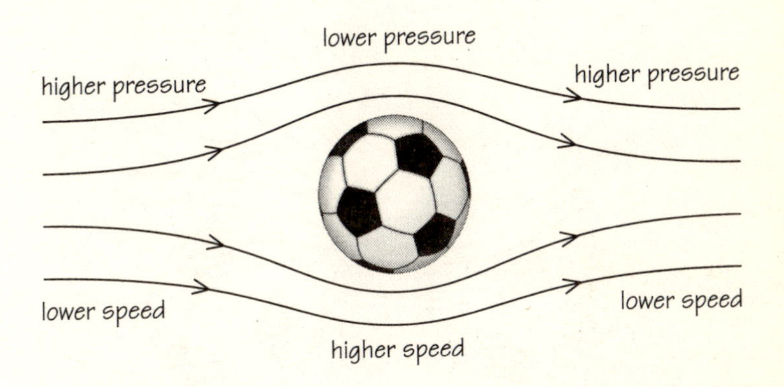

If the ball is spinning then the flow lines of the air near the surface of the ball are significantly altered. In the picture below you can see what happens if the ball has been given a clockwise spin. At the top the air movement very close to the surface of the ball is in the opposite direction to the oncoming air whereas at the bottom it is in the same direction. This means that the net speed of the air near the top of the ball is less than that near the bottom. Therefore the pressure on the ball is greater at the top than the bottom and there is a net downward force. The orientation of the picture shows how a ball with topspin will swerve downwards. If we look from above it describes how a ball struck with the outside of the right foot will swerve to the right.

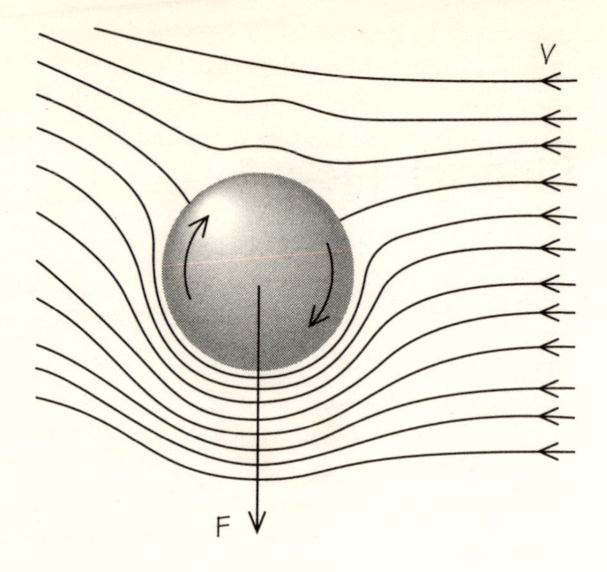

During the last football World Cup in South Africa there was considerable controversy and much complaining by goalkeepers about the introduction of a new lighter ball that displayed unfamiliar aerodynamic properties. Yet it was very noticeable that the world's leading attacking players never really mastered the behaviour of the ball and there were almost no goals scored by long-range shots or free kicks. Players were simply unable to control the swerve of the ball.

91

Stop-go Tactics

If a runner, a cyclist or a motorist makes many short stops to brake at traffic lights and junctions or, in the case of runners, perhaps as part of a stop–go sequence of fast training runs, where does all their energy go? For the driver and the cyclist the energy of motion ends up being dissipated in the brakes where it is dumped into heat and sound. If the runner stops suddenly his energy is lost in stretching and heating the muscles, tendons and limbs which act as a physiological braking system.

When our runners and riders are not stopping and starting they are working to overcome the resistance provided by the air they move through. It is interesting to see when the stopping and starting is a bigger drain on their energies than the air's resistance. If each mover has a mass M and proceeds at a constant speed V between a series of sudden stops a distance D apart, then the time between stops is D/V and all the kinetic energy, $\frac{1}{2}MV^2$, is lost in the 'brakes' at each stop.[1] This means that the rate of loss of energy into the braking system is $\frac{1}{2}MV^2 \div D/V = \frac{1}{2}MV^3/D$. The shorter the stretches between stops (the smaller D is) the larger this becomes. It also increases like the cube of the speed, so moving fast means there will be a lot of energy wastage when you stop.

The other drain on the runners' and riders' energies is overcoming air resistance to their motion. If they present an effective area[2] A towards the air, then over a time t they sweep out a cylinder of air that has a volume equal to A times the distance travelled; this is $V \times t$. This cylinder therefore contains a mass of air equal to $M_{air} = \rho AVt$, where $\rho = 1.3kg/m^3$ is the density of the air. The kinetic energy of this displaced swirling air will be $\frac{1}{2}M_{air}V^2 = \frac{1}{2}\rho AtV^3$ and the rate of energy loss in overcoming it is $\frac{1}{2}M_{air}V^2/t = \frac{1}{2}\rho AV^3$. If we compare this with the braking losses we see that:

$$\text{Energy loss rate to braking/Energy loss rate to air} = M/\rho AD$$

What this shows us is that if M is bigger than ρAD, which is the mass of air in that cylinder being swept out by the moving runner or rider, then more energy is wasted in repeated braking than in overcoming the drag of the air. This condition makes simple sense. You will find your energy losses dominated by repeated stops if your mass, M, is large because that means there is more kinetic energy to dissipate in the brakes (it's harder work to stop a huge truck than a bicycle) or if D, the distance between stops is small, so you are making lots of them. When M equals ρAD it picks out a special distance between stops, $D^* = M/\rho A$. This separates the situation where the repeated brakings dissipate most energy ($D < D^*$) from the opposite case where the air drag is the biggest drain on energy ($D > D^*$). For a distance runner, the effective area is $A = 0.45m^2$ and $M = 65kg$, so $D^*(\text{runner}) = 111m$. For a racing cyclist and bike, $A = 0.25m^2$ and $M = 75kg$ so $D^*(\text{cyclist}) = 230m$, whereas for a typical car $A = 1m^2$ and $M = 1{,}000kg$, so $D^*(\text{car}) = 750m$. For the motorist you can see that when the stops between junctions and traffic lights are less than 750m the braking losses dominate and in order to save energy you should drive a lightweight car, and drive slower because the energy losses depend on mV^3. For the runner, the aim in training may be the opposite of energy saving. If you are doing a series of fast training runs with short stops between them and if the runs are each less than about 111m in length, then you will be using up more energy stopping and starting than cruising through the wind. By reducing the interval a little further you can use even more energy and this will have a greater beneficial training effect than running more slowly over longer distances.

92

Diving is a Gas

Scuba-diving is a sport that must be unique because it requires knowledge of the gas laws of physical chemistry that you should have learnt at school.[1] This understanding forms part of the basic proficiency certification needed to take part in organised diving.

The first and most important gas law was investigated extensively by Sir Robert Boyle in the seventeenth century.[2] It reveals that the pressure of a confined gas is inversely proportional to its volume: double its pressure and you will halve its volume. 'Boyle's Law' illustrates what happens when a diver descends and the water pressure on his body and wetsuit rises. Atmospheric pressure at sea level is defined as 1 bar.[3] A diver will experience an increase of this by an extra 1 bar for each 10m of descent underwater. Any trapped air is compressed into a smaller volume and so his buoyancy suit deflates a little and his wetsuit no longer fits as tightly as it did on the surface. More significantly, the air in the Eustachian tubes between the ears is compressed by the increasing pressure and you need to push some air into them ('equalise') to compensate for this. This is that familiar feeling of making your ears 'pop'. When you go back up to the water surface, gases in and around your body will expand as you ascend to lower water pressures. If you made the mistake of holding your breath, instead of gently exhaling, then the air in your lungs will expand and damage (or even burst) your lungs.

The next important gas law to know is 'Henry's Law', discovered by the Mancunian chemist William Henry in 1801. It is the dictum that the mass of a gas that dissolves in a liquid is proportional to the pressure of that gas. This means that a diver's body will absorb more gases into the bloodstream and tissue as the pressure on the body rises at increasing depth. Much larger amounts of nitrogen will be absorbed from the air tanks and this needs to be carefully managed when resurfacing. When you climb out of the water from a lengthy dive you will have much more nitrogen in your system that normal. The deeper you went, and the longer you spent underwater, so the more you will have retained. It will gradually be expelled by the body but a careful check is kept on the length of dives and the time between them in order to keep this gas retention within safe limits.

Gas absorption by the diver's body is also subject to Dalton's Law, introduced by another Mancunian, John Dalton, in 1803. It requires that mixed gases remain in the same proportions when their total pressure changes – that is, in a mixture of oxygen and nitrogen the two gases don't respond differently to changing pressure like a non-gaseous mixture of, say, grapes and marbles would. Dalton's Law allows us to predict and monitor the body's absorption of different gases in the breathing mixture fairly straightforwardly.

These gas laws predict that there is far greater danger to divers when they come up to the surface than when they descend. Any additional nitrogen and oxygen dissolved in the blood can form tiny bubbles in the arteries if a diver ascends too quickly into water at lower pressure. This can lead to pains in the joints ('the bends'), aches in teeth and dental fillings, and the formation of potentially fatal embolisms as a result of migrating and coalescing gas bubbles in the arteries. It is important to ascend slowly from deep dives (less than 10m per minute) to gradually reduce the pressure of gases dissolved in the body and allow time for out-gassing to occur. Experienced divers monitor this very carefully and, in the event of serious problems, decompression chambers are available to provide an artificial high-pressure environment to allow divers to undergo a slower return to sea-level pressures.

93

Spring is in the Air

When you give something a gentle push so that it wobbles before returning to its previous undisturbed state then it will oscillate back and forth at a particular 'natural' frequency. If you swing back and forth on a children's swing and don't pump your arms and legs, then you feel this natural frequency for the swing. If you start to force the oscillations of the swing to go higher by pulling on the swing cords and moving your legs back and forth in unison then you can enhance the amplitude of the swings you make. You soon realise that there is a right and a wrong way to do this. If you push at the wrong moments it seems to have a negative effect but if you push at just the right moments to reinforce the natural motion then you produce a greatly enhanced upward swing.

The key for success is to apply your forcing at the same frequency as the natural frequency of the oscillating swing, so that the time between pushes is the time between oscillations at the swing's natural frequency. This leads to very efficient energy transfer into the motion of the swing and is called 'resonance'. Sometimes resonance is good, as for a child on the swing; but sometimes, for instance when an earth tremor shakes a house, resonances are bad and need to be avoided by good engineering that filters out the dangerous frequencies.

The most dramatic example of resonance in sports is to be found in springboard diving. The type of flexible diving board employed is usually 3m above the water (perhaps only 1m in a recreational pool) and is made from a single piece of very strong aluminium,[1] about 4.9m long and 50cm wide, with a non-slip epoxy coating to give the divers a sure footing.

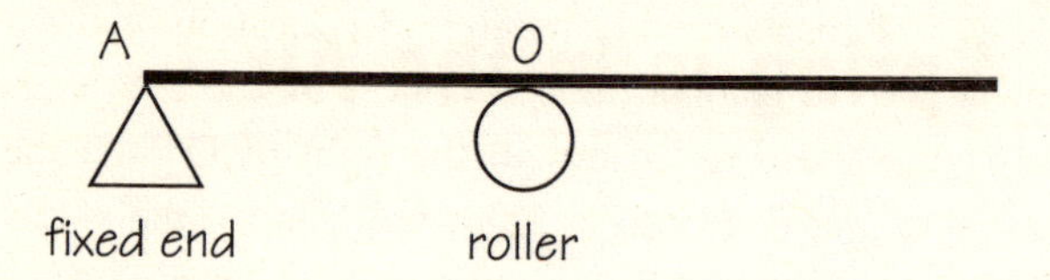

The board is fixed at one end (A) and free to oscillate at the other, which overhangs the pool. This is what engineers call a 'cantilever'. However, the springboard is set up like an adjustable lever with a fulcrum at O. The diver can adjust the position of the fulcrum to within about 0.75m by rolling a gear wheel with his foot when he stands near A before commencing the dive. Typically, the diver will take three well balanced steps along the board so that the last of them takes place about 1m beyond the fulcrum. The third step leads immediately into the launch, with the help of the board's upward motion. When the diver jumps up – and also a little forwards so as to clear the board on the way down) he raises one leg along with his arms; this produces an equal and opposite downward reaction on the board. It bends the board slightly downwards before springing back towards its undisturbed state with a force kx, where x is the distance it was deflected and k is the springiness of the board.[2] When the board is made to oscillate gently it will do so at its natural frequency, which is $\surd(k/m) \sim \surd 848/65 = 3.6$ per second for a 65kg diver. This is the frequency that the diver needs to tune into when he bounces on the board, so that he receives an especially large 'resonant' upward impulse when he parts contact with the board. It amounts to hitting the board on the last contact when the board is moving downwards at its highest speed so that you maximise its downward deflection and get most of that stored elastic energy transferred into your upward launch.

A resonance between each of his bouncing movements on the board and the board's natural oscillation frequency transfers energy most efficiently into his upward motion. The adjustments of up to 0.8–1.2m he is allowed to make to the location of the fulcrum O enable him to change the springiness of the board. If he has more of the board rolled out to the right of O then the board will be springier (larger k) and this will change the value of the resonant frequency that he is aiming to catch because it is proportional to the square root of k. Divers of different individual sizes and strengths will choose to use different fulcrum positions so that they can most effectively force the board to oscillate at the resonant frequency for their mass. When a diver hits the resonant frequency of the board just right, you can hear it. There is a nice single twang of the board as the diver is launched into the air. Miss the right frequency and there is a clatter of vibrations for all to hear.

94

The Toss of the Coin

Tossing a coin is the answer to all sorts of problems in sporting events – who kicks off, who has which end of the ground, who has the wind at their back, who serves first, and even who qualifies for the next round or wins the match when all other ways of choosing based on merit have been exhausted. The reason is that this simple device is perceived to be completely random and unbiased with an equal probability of falling with heads or tails facing upwards.

In some sense the tossing of a coin cannot be truly random.[1] There is a definite starting state – heads facing upwards, for example – and the coin is thrown upwards with some spin. It follows a trajectory determined by the force of gravity and returns to be caught by the thrower after a definite number of turns. All of this is described by Newton's laws of motion. If we launch the coin upwards with a vertical speed V from a height H_0 above the ground then it will rise to a height $H = H_0 + Vt - \frac{1}{2}gt^2$ after a time t, where g is the acceleration due to gravity. It will return to the thrower's hand at the same height, H_0, where $H = H_0$, again after a time $t_h = 2V/g$. If it was also tossed upwards with a spin of R revolutions per second (rps) then it will have turned over completely N times where:

$$N = t_h \times R = 2VR/g$$

You can see from this formula that if you want the number of coin turnovers to be large then the upward launch speed, V, should be large so it travels a nice long distance in the air. The spin on the coin is, of course, essential. If you throw it up with very little or no spin, like a Frisbee, it won't turn over and will land with the same face up as you started with. The formula also reveals the degree of predictability. If N was only 1 and the coin was launched very slowly, with heads upwards, then it will fall with heads up again. When N is between 2 and 3, or 4 and 5, or 6 and 7, and so on, it will be caught with the same face up as it was thrown with, but if N lies between 3 and 4, or 5 and 6, or 7 and 8, and so on, then it will land with the opposite face upwards.[2] When N gets large, much bigger than 20, say, the conditions on V and R that distinguish the two outcomes get closer and closer together and very small differences in the tossing conditions result in heads or tails. Typically, V will be about 2m/s and with $g = 9.8$ m/s^2, and so we get a time in the air of $2V/g = 0.4s$. In order to have time for more than 20 revolutions of the coin so as to make the outcome close to 50–50, you will need to spin the coin at a rate in excess of $20/0.4 = 50$rps.

The key to making the coin toss fair for the two captains who might be calling 'heads' or 'tails' is the setting up of the initial state of the coin. If the referee hides the face from the captains then they will not be able to gain an advantage from being the one to make the call: if they don't see the initial upward face, they won't know of any bias that makes the coin preferentially fall 'heads', say. For a good coin toss we have just seen that throwing it high in the air gives lots of turnovers. It spends half its time in the air with 'heads' upwards and 'tails' downwards and this results in the fraction of outcomes being 'heads' when it is caught by the referee getting very close to 1 in 2.

It is interesting to note that you can't actually bias the coin by changing the weight distribution. Concerns were voiced in 2002 that Belgian euro coins had a significant weight bias to the 'heads' side, which featured King Albert II. Fortunately, there was nothing to worry about when it came to tossing Belgian euros at the start of football games. Making one side heavier than the other doesn't create a bias that matters: the coin will always spin through the air about an axis that passes through its centre of gravity, no matter how lopsided it is.

95

What Sports Should Be in the Olympics?

The International Olympic Committee (IOC) has a lengthy list of criteria that it takes into consideration when deciding which sports qualify to be considered for inclusion in (or removal from) the hallowed roster of Olympic sports or the waiting room that holds the 'demonstration' sports, which aspire to attract the most IOC members' votes for inclusion. At present, there are twenty-six sports in the Summer Games but this will increase to twenty-eight, the maximum allowed by the IOC rules, when golf and rugby sevens are included in 2016 and 2020.

The process of considering which sports could be added involves a questionnaire to IOC members that focuses on seven aspects of a sport: history, universality, popularity, image, athletes' health, development of the international federation and costs.

All of these are relevant considerations but they don't help much in winnowing down the bloated package of Olympic sports currently on the roster. There is one absent consideration that seems to me to be important – and while meeting it is not a sufficient condition for inclusion in the Games, it should be a necessary one. It is to ask whether winning the Olympic Games is the pinnacle of sporting achievement in that discipline. This is manifestly the case in athletics,[1] swimming, track cycling, hockey, volleyball, table tennis and almost all the other sports on the roster. It would also be the case in sports like karate and squash that are seeking to become Olympic sports. However, there are glaring examples where it is not the case. Tennis, golf, football, basketball and the former Olympic sport of baseball, all fail this litmus test. Olympic football is particularly odd in that it actually limits teams to only three professional players older than 23 years of age; no other sport has this artificial restriction on a team's strength. Moreover, top competitors in Olympic tennis, golf, football and basketball have other primary career goals and many choose not to participate. Would you rather win Wimbledon or the Olympics? The Olympic football tournament or the World Cup? The answers are obvious and should be a key factor in determining whether these sports are suitable for inclusion in future Games.

96

The Cat Paradox

When you watch divers and trampolinists they seem to defy the laws of mechanics. They start by moving downwards or upwards without any rotation but then they can perform a sequence of somersaults and twists. How is this possible? There is nothing for them to push on to create a torque that will make them twist. In fact the rules of diving actually forbid twisting from occurring directly from the diving board or platform.[1] When objects rotate there is a measure of the rotation, called the angular momentum[2] that must be conserved in the process. It means that you can't spontaneously create overall net rotation. Yet, top high-board divers perform spectacular twists, and even manage to keep their body straight in the process. How do they do it?

The same technique for creating twist is used by falling cats. Whereas divers want to twist to score points and enter the water head first in a vertical position, cats generally don't care about the points; they just want to land on their feet. They also have two attributes that divers don't – a special bone structure and a reflex which distinguishes 'up' from 'down'. When they fall from a height, cats first bend in the middle so they can pull in their front legs and extend their rear legs. This decreases the inertia in the front part of their body while increasing it in the rear half. This enables the front half to rotate quicker and a lot further than the back half does in the opposite direction during the same time. The front and rear parts of the body are rotating about different axes in different directions but when the two are added together, one positive and one negative because of the different directions, the sum is zero just as it was at the outset of the fall. Next, they extend their front legs and pull in their back legs which allows the back half of the body to rotate a lot while the front half rotates far less in the opposite direction. If necessary, the cat can repeat these steps quickly so as to get in just the right position for a soft landing (see figure below).

The high-board diver plays the same trick. He leaves the diving board with zero spin and angular momentum. He moves one arm upwards and one downwards below his chest so they are both rotating clockwise while the rest of his body rotates anticlockwise (see figure below). The frictional force on his foot when he leaves the board provides a sideways torque that ensures his body starts to twist as soon as it begins to rotate. Eventually, he stops the twisting by stretching his arms out and straightening his body so his inertia rises and the spin drops. It sounds easy but it certainly isn't. The cat makes it look very easy though, and it is worth watching next time you see a cat dropping off the banisters. Apparently, some have even survived falls from the window ledges of high-rise New York skyscrapers.

97

Things That Fly Through the Air With the Greatest of Ease

Countless sports involve throwing, kicking or batting small objects – balls or shuttlecocks – through the air. Some balls are made of leather, some are plastic; some are big, some are small; some are heavy, some are light. Yet, despite the diversity, any old ball will not do. Soccer balls mustn't be too light and bouncy or they will bounce over the goal at the other end of the pitch when the goalkeeper kicks it high and hard downfield. Table-tennis balls mustn't be too heavy or they won't move quickly or deviously enough. Can we make some sense of the array of different sporting projectiles and see if they have been chosen so as to make their game 'interesting' in some clear sense?

Once you launch a sports ball through the air it faces two decelerating forces: gravity due to its weight, $W = mg$, and the drag force, $D = \frac{1}{2}CA\rho V^2$, imposed by the air it moves through, where m is its mass, g the acceleration due to gravity, A the surface area at right angles to the direction of motion, V its speed relative to the air, ρ the air density, and C the drag factor determined by the ball's smoothness and other aerodynamic properties. The relative importance of these two forces will be given by their ratio:

$$\text{Drag/Weight} = CA\rho V^2/2mg$$

The second important factor in determining the nature of the ball's motion is the relative importance of the drag force and the friction, or viscosity, of the medium. This will determine the division between flows which are smooth and orderly and those which are turbulent and chaotic. The ratio of these two forces, $\rho V^2 A / \rho \nu L$, where $\nu = 1.5 \times 10^{-5} m^2/s$ for air at 20°C is the viscosity of the medium and L is a length which characterises the scale of the flow (it might be the circumference of a ball, for example), is called the 'Reynolds number' of the motion and is denoted by *Re*. It is a pure number and has no units (because it's a force divided by a force). When *Re* is large the flow passing very close to the surface of the ball will become turbulent but when it is small the flow there will remain smooth. The transition between the two states can be quite abrupt and will produce subtle behaviour as the projectile moves through the air, with small changes in spin, air density or ball texture having marked effects. This 'interesting' transition value for the Reynolds number is typically about $Re = 1$ to 2×10^5.

In the table[1] below we have given the defining masses and dimensions for a range of projectiles in sport, together with the numerical values of the drag-to-weight ratio (D/W) and the Reynolds number. The range of values for the drag-to-weight ratio is quite large: from 0.01 for the shot to 8.8 for a table-tennis ball. However, despite this diversity and the different nature of the sports we have tabulated, it is very striking that the Reynolds numbers for all of them are rather similar. Our ball games have homed in on this interesting situation where the ball does subtle things when a player provides a little spin or variation in the way it is struck. This is one of the features that make ball games interesting for players and spectators alike, as unpredictable motion of the new footballs used at the 2010 World Cup revealed.

Sports projectile	A(m^2)	m(kg)	V(m/s)	L(m)	D/W	*Re*
Tennis ball	0.003	0.06	40	0.06	1.00	1.6×10^5
Table-tennis ball	0.001	0.002	25	0.04	8.8	0.7×10^5
Squash ball	0.001	0.02	50	0.04	3.52	1.3×10^5
Soccer ball	0.038	0.42	30	0.22	1.02	4.4×10^5
Rugby ball	0.028	0.42	30	0.19	0.75	3.8×10^5
Basketball	0.045	0.62	15	0.24	0.2	2.4×10^5
Golf ball	0.001	0.05	70	0.04	1.23	1.9×10^5
Shuttlecock	0.001	0.005	35	0.04	27	0.2×10^5
Shot-put (men)	0.011	7.26	15	0.12	0.01	1.1×10^5
Water polo ball	0.038	0.42	15	0.22	0.23	2.2×10^5
Javelin (men)	0.001	0.8	30	0.03	0.64	0.6×10^5

98

Some Like It Hot

The new velodrome for the Olympic track cycling has attracted a lot of attention because of its optimally engineered seamless gradients and atmospheric surround seating. Both features will add to the atmosphere and speed of the performances on the track. This extra special care may not be unconnected to the fact that cycling has been Britain's most successful sport at recent games and there are great expectations that the British team can do even better in front of a home crowd.

There is another feature of the new velodrome that is likely to be a big talking point during the games: its temperature. The air temperature at track level will be artificially kept considerably higher than usual, up from about 20°C to 25°C or more; although we are assured that the spectators' area will be cooled to normal levels. This doesn't sound appealing if you are sweating your guts out racing the pursuit for 4km, so what's behind it? It's our old friend air drag again. The air resistance on the cyclists is proportional to the density of the air that they are moving through and that density depends on the air's temperature.

The density of air falls as the temperature rises. At the molecular level, the rising temperature makes the molecules jiggle about with a faster average speed and the average volume they occupy increases, and so the density (the mass of molecules divided by the volume occupied) of those moving molecules decreases. The variation is shown in the first figure.[1]

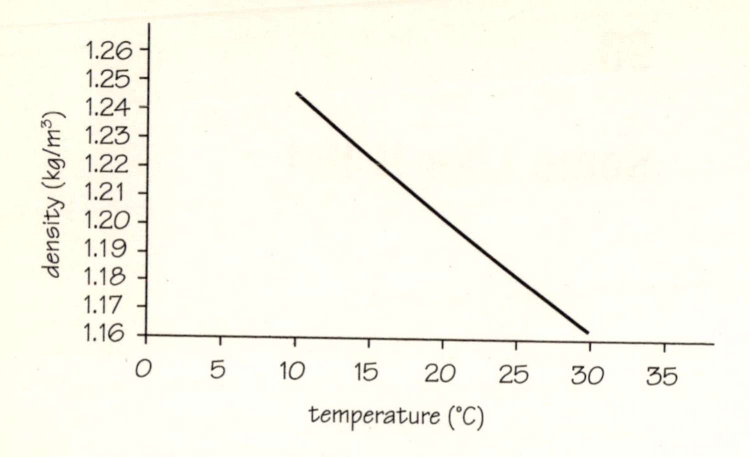

The expected reduction in drag over a range of temperatures that results from this is shown in the second figure. The drag increases like the square of a cyclist's speed and the drag reduction created by raising the air temperature will improve times by about 1.5s over the 4km pursuit distance. Unlike finding improvements in bike aerodynamics or clothing, this feature of the velodrome helps all competitors equally (although the fastest have more to gain than the slowest) and will lead to an increased chance of world records being set on the track in all events. However, this situation is not entirely satisfactory. It is reminiscent of the Mexico Olympics of 1968 where a reduction in air drag was created by the lower air density at the high altitude of 2,240m in Mexico City. As we have seen, this distorted performances in the track athletics and horizontal jumps so much that an 'altitude-assisted' proviso had to be attached to the 'records' set there and at subsequent high-altitude venues. It seems that we are artificially recreating this two-track situation in cycling with any future records having to have the air temperature at track level attached to them. Those performances registered at too high a temperature will end up being sidelined as 'heat-assisted' just like those altitude-assisted track performances.

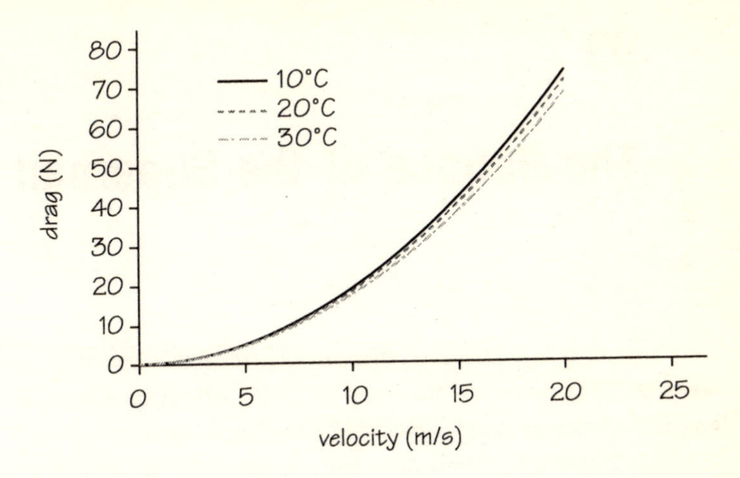
10°C
20°C
30°C
drag (N)
80
70
60
50
40
30
20
10
0
0
5
10
15
20
25
velocity (m/s)

99

The Bounce of the Superball

The 'Superball' has fascinated children and threatened greenhouse windows ever since its invention in 1965, when the chemical engineer Norman Stingley patented it under the prosaic name of a 'Highly Resilient Polybutadiene Ball'. In fact, Stingley had discovered it accidentally, in the form of a plastic that bounced uncontrollably. It was soon renamed, mass-produced, and sold by the million in the 1960s for just under a dollar a time by the Californian Wham-O company. Its secret was the vulcanised synthetic rubber polymer used to manufacture it. This gives it a very high coefficient of restitution, as physicists would say.[1] In short, when it is dropped (not thrown) from a height it bounces back to more than 90% of the height it started out from. If you throw it hard it can bounce over your house!

The Superball has one great claim to sporting fame. In the late 1960s Lamar Hunt, the founder of the American Football League, coined the name 'Super Bowl' for American football's cup final after watching his children play so enthusiastically with their new Superball.

The Superball does more than bounce high. It bounces in a very unusual and counter-intuitive way because its surface is almost perfectly rough. Its behaviour can appear very mysterious to observers because it doesn't behave like a tennis ball or a billiard ball when it impacts with the ground. This odd behaviour is why it can be very hard to catch the ball once it bounces on the ground – you don't know which direction it will rebound. The first figure[2] shows what happens when a Superball is bounced between the floor and the underside of a table. The dashed line shows the expected trajectory of a conventional ball. If it is thrown forwards it keeps moving forwards. But a spinning Superball that starts moving forwards can reverse its direction of motion at a bounce and come straight back towards you!

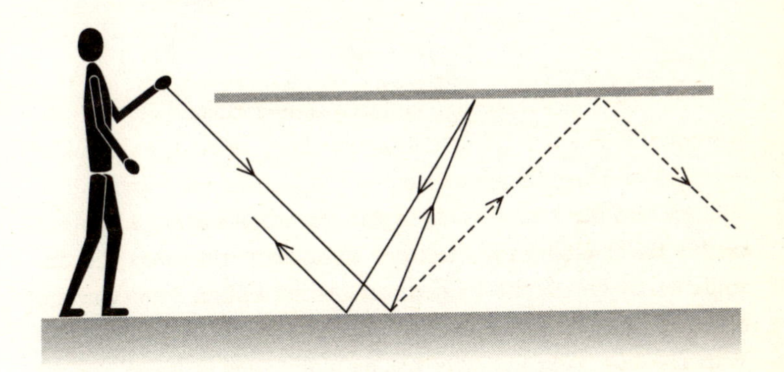

This behaviour arises because the rough Superball does not slip when it makes contact with a surface and the energy of motion is conserved at each bounce, aside from small losses due to heat and sound (which we ignore). If the ball is spinning then the energy of motion has two parts: the kinetic energy of the centre of mass of the ball and its rotational energy about that centre. At each bounce it also conserves angular momentum and these two principles allow the motion to be predicted.[3] It is possible to see what happens by using some simple symmetry principles rather than solving the complicated equations.[4] Imagine that the Superball is heading towards the ground and has some backspin (see the second figure) and loses no vertical velocity when it bounces. The question is: what will it do after a bounce?

The bounces have to obey a fundamental principle of Newtonian motion: they must be reversible in time.[5] That is, if a ball moving in a state 1 bounces to be in state 2, then the time-reverse of state 2 must bounce to become the time-reverse of state 1.

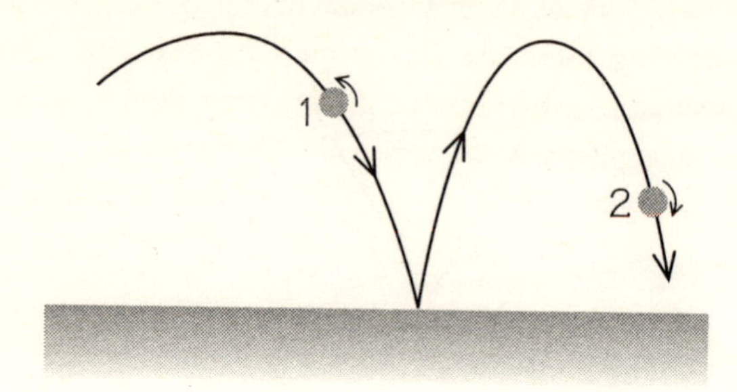

We can find the time-reverse of the Superball's state of motion simply by reversing the direction of motion and reversing the sense of its spin. If we look at our second figure, we see that a Superball has started with backspin and bounced into a state with topspin. When it bounces a second time it must end up in the same state in which it began. If we reverse the direction of motion and of spin at the second bounce, then the ball will follow the same parabolic path as it did in state 1, moving from left to right with backspin.

When we throw the ball without any spin so as to bounce between the floor and the underside of the table, as in our first figure, then the ball follows the trajectory shown, and returns after three bounces towards the thrower, with slightly reduced speed. This is possible because of the uniform mass distribution within the Superball. If there was a dense mass concentration near the centre of the ball,[6] it could have exactly reversed its path after two bounces, as it does in the next figure below.

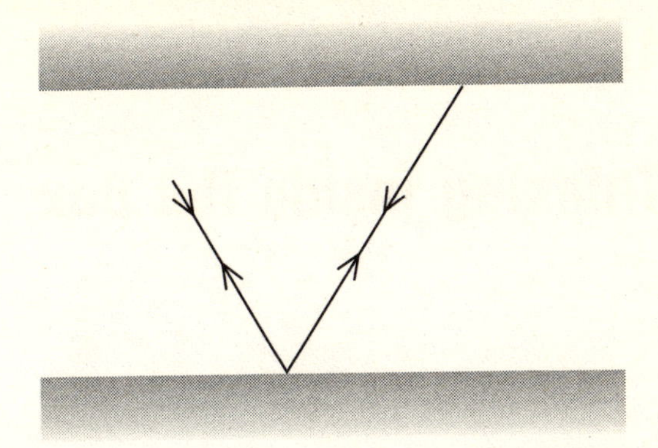

Finally, playing snooker with a Superball is an interesting experience that will tax the finest player. The figure below shows what happens to a smooth snooker ball after it is struck at 45 degrees to the side of a square table[7] so as to follow a closed path inside the square (right) compared to the trajectory of a rough Superball that is given the same impulse (left).

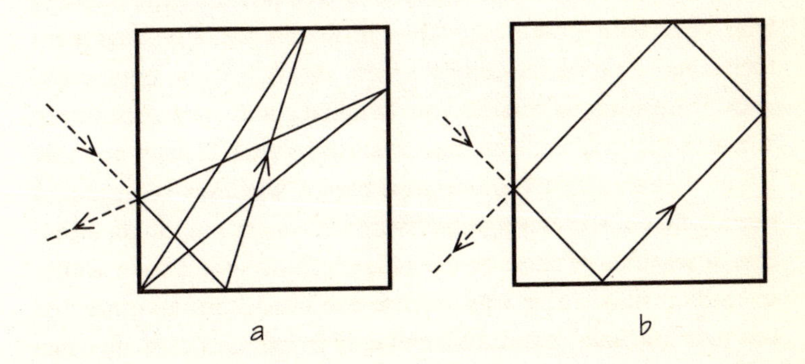

A little practice on the snooker table with a Superball will soon build up your intuition about the behavioural differences of smooth and rough balls – but don't strike it too hard.

100

Thinking Inside the Box

After my book *100 Essential Things You Didn't Know You Didn't Know* was published, I received lots of emails and letters asking about some of the mathematical applications and issues that appeared there. One topic was overwhelmingly the most common. Some people didn't understand it; others just didn't believe it. Some puzzled correspondents were just interested readers, but at least one was a very famous professor of physics. What worried them so much was the infamous 'Three-box', or 'Monty Hall', problem in chapter 30. You may remember how it goes. There are three boxes. One contains a prize and you have to pick the prize-winning box. The game-show host, who can see all the boxes' contents, then opens one of the empty boxes, shows it to be empty, and asks if you want to keep to your original choice or switch boxes. What should you do? You should always switch. There was only a 1 in 3 chance that your original choice was right but a 2 in 3 chance that it was wrong and the prize was in one of the other two boxes. One of those two was opened and shown to be empty so there is now a 2 in 3 probability the prize is in the other box you didn't choose – switch to it and double your chance of winning!

This problem is not paradoxical – just counter-intuitive – but it's a good warm-up for another problem that is much more perplexing. This time we play a game in which there are just two boxes and in one of them there will be a prize of some value and in the other another prize of twice its value. Pick one of the boxes and open it. You are now offered the chance to switch your choice to the other box. What should you do? The problem is that you don't know if you opened the box containing the smaller or the larger prize. So let's see what the expected pay-off is if you switch after opening one box to find it contains a prize of value V. If it was the smaller prize, then by switching you would get 2V, but if it was the larger prize you would end up with ½V. So, on average, a switch will bring you an expected prize value of $(\frac{1}{2} \times 2V) + (\frac{1}{2} \times \frac{1}{2}V) = 1\frac{1}{4}V$. This is bigger than V so you *always* gain on the average by switching.

This is very odd because if you *always* gain by switching you didn't need to look in the box did you? Always switch. And after you've switched, the same argument applies and you should switch again! This is a paradox.[1] What do you make of it? An interesting clue is to start by asking what would happen if the prizes were actually £5 and £10 (or $5 and $10) banknotes.

Notes

Chapter 1

1 M. W. Denny, *J. Exp. Biol.* 211, 3836 (2008). The predictions in Denny's article have received quite widespread publicity in the *New Scientist* and other publications. Unfortunately, however, the analysis in this paper ignored the role of reaction times and predicted a maximum future running speed for Bolt that is slower than what he is running now.

2 Top athletes take about 0.3s to get out of the blocks.

3 J. D. Barrow, 'Frankie's Fastest', *Athletics Weekly* magazine, 6 August 1997.

4 J. R. Mureika, New Studies in Athletics 15 (3/4), 53–60 (2000).

5 Ironically, Bolt responded a little too quickly at the 2011 World Championships in Daegu and was disqualified for false-starting. Remarkably, his teammate Yohan Blake, who won the 200m in Bolt's absence, went on to run in 19.26s (with a +0.7m/s wind), only 0.07s slower than Bolt's world record. More remarkable still, Blake's reaction time was a very lethargic 0.269s, whereas Bolt's was 0.133s for his 19.19s world record in Berlin in 2009. Thus we see that Blake's 200m running time was 18.99s against 19.06s for Bolt.

Chapter 3

1 This record has recently been improved to 693 points in October 2011 by the Korean Im Dong-hyun. This is especially remarkable because Im has 20/200 vision and so is 'legally' blind. He uses special glasses or contact lenses in competition.

2 The score per dart is 12.8 but we have rounded to the nearest whole number. For some hints about this calculation, and the areas of the different dartboard sectors, see R. Eastaway and J. Haigh, *How to Take a Penalty*, Robson Books, London (2005), pp66–8.

Chapter 5

1 If the radius of curvature of lane number L from the inside is r, then $r = 100/\pi + 1.5(L - 1)$ m.

2 A detailed study of the equation of motion for a runner sprinting around a curve has been attempted by J. Mureika, Can. J. Phys. 75, 837 (1997) where more details can be found. The sprinter is approximated by the solution of the non-linear differential equation introduced by R. Tibshirani, *Amer. Statistician* 51, 106 (1997), as an extension of the old model of Hill and Keller, of the form $dv/dt = -v/T + [(f - st)^2 - k^2v^4/R^2]^{1/2}$ where T is a physiological factor, k is the fraction of the full centrifugal force that the runner feels around the curve, f is a fatigue factor and s is a constant.

3 This was hand-timed which always gives slightly faster timing than electric timing to one hundredth of a second. However, there is another interesting subtlety about the timing of Smith's straight 200m. The timekeepers are a very long way away from the starter and have to see the smoke (not hear the sound which travels slower) from his pistol in order to start their watches. The farther you are away the slower you respond and the faster the time you end up recording, This effect has been estimated to quicken recorded hand times by as much as 0.14–0.24s for straight 200m races. Modern races have fully automatic electric timing and there are no timekeeping effects of this sort any more in top-class races.

Chapter 6

1 The wobble time for a complete oscillation will be proportional to the square root of your inertia. Since your inertia is proportional to your mass and the square of the dimension over which you have spread your mass, this means that the wobble time will be proportional to the arm-spread distance – the bigger it is the slower the wobble.

Chapter 7

1 This unusual distance is called a 'chain' in England and dates from the 1600s when it was introduced as a unit of measurement in land surveying. Originally, it referred to an actual measuring chain of links that was used like a tape measure.

Chapter 9

1 If the probability of scoring a goal is independent of scoring or conceding an earlier or a later goal, as assumed here, the probability of several goals is just the product of the probability of each.

2 Much less than ½ so we neglect s^2 and s^3 compared with s when calculating p^3.

Chapter 11

1 The birthday distributions among elite French footballers were studied in N. Delorme, J. Boiché and M. Raspaud, *Eur. J. Sport Sci.* 10, 91–6 (2010) and found to manifest a bias to the first two quarters of the year. The skewed distribution of birthdates has also been observed in baseball, cricket, tennis and rugby in a range of cultures. These studies are reviewed in J. Musch and S. Grondin, *Developmental Review* 21, 147–67 (2001).

Chapter 12

1 This kick is named after an Irish rugby club in County Limerick.

2 T. D. Lipscombe, *The Physics of Rugby*, Univ. Nottingham Press, Nottingham (2009).

Chapter 13

1 There will also be a 200m event at the 2012 Games.

Chapter 14

1 Olympic events take place over a straight-line course and so the cox doesn't have the huge tactical importance that he or she has in a race like the Oxford vs Cambridge boat race over four miles of the River Thames with long bends, difficult currents and many choices to be made about the course to be taken or the stroke rate.

2 If we hadn't allowed for the smaller size of the cox we would have got ratios of 1.09 and 1.05.

Chapter 16

1 The rider will also need to supply energy to overcome air resistance and friction with the road.

2 The total mass $M = m_{frame} + 2m$ where m_{frame} is the mass of the bike frame plus the rider. We are neglecting the energy required to rotate the pedals of the bike but the gearing of the pedal rotation circle compared to the wheel radius ensures that the pedal speed is only

about one fifth of the wheel's speed and so the energy needed to spin the pedals is twenty-five times less than the energy needed to move the wheel at speed v and so can be ignored as small.

Chapter 17

1 See http://www.iaaf.org/mm/Document/Competitions/Technical Area/ScoringTables_CE_744 .pdf.

Chapter 19

1 $P = mdV/dt \times V = ½d(V^2)/dt$ so $V^2 = 2Pt$ if $V = 0$ at $t = 0$. To find the distance x travelled in time t, use $V = dx/dt$ and so $x = (8P/9)^{3/2}t^{3/2}$.
2 G. T. Fox, *Am. J. Phys.* 41, 311 (1973).
3 We have ignored the short period of rolling at the start where the traction is limited by friction. During this phase the kinetic energy of the car equals the work done against the frictional resistance (which equals the friction coefficient μ times the weight of the car, mg, times the distance gone, x,) so $½mV^2 = \mu mgx$. It therefore continues until a value of the velocity $V = V_0 = (2\mu gx_0)^{1/2}$ is reached that corresponds to the constant power relation $V = (3Px/m)^{1/3}$ and then it moves at constant power for $x \geq x_0$. For more details see the article by R. Stephenson, *Am. J. Phys.* 50, 1150 (1982). It is also possible to take into account the loss of fuel mass with time.

Chapter 20

1 Javelin throwers and high jumpers wear shoes that have spikes on the heel as well as the sole in order to prevent slip.
2 The frictional force opposing horizontal motion is μW where W is the weight of the body and μ is called the coefficient of static friction. Static friction does not depend on the area of mutual contact between the two surfaces.
3 These are values for static frictional resistance. Once motion begins, the frictional resistance to continuing motion is slightly smaller. The frictional resistance to rolling motion, say by a wheel, is different and is proportional to the weight divided by the radius of the rolling wheel.

Chapter 21

1 This is clay pigeon shooting with a shotgun. There is a random delay time between 0 and 3 seconds after the shooter has called for the moving clay to be released.

2 This made Murdock the first woman to win an Olympic shooting medal. She tied for the gold with the US team captain, Lanny Bassham, but no shoot-out was allowed under the rules to decide the gold. Instead, the best performance from the last ten shots was used to break the tie. Most shooters regarded this as an unfair rule and Bassham and others wanted two gold medals awarded if there couldn't be a shoot-out. But this request was turned down by the Olympic committee. Finally, at the medal ceremony, Bassham pulled Murdock up on to the gold medal position on the podium with him before the anthem was played.

Chapter 23

1 Roller coasters are not circular; otherwise the force exerted on riders at the bottom of the ride would be dangerously large. The curve used is a 'clothoid'. See John D. Barrow, *100 Essential Things You Didn't Know You Didn't Know*, Bodley Head, London (2008), chap. 50.
2 For another analysis of the safety of this exercise, see M. S. Townend, 'Modelling the Giant Swing – Is It Safe?', *Teaching Maths. Appl.* 12, 163 (1993). For further technical literature on the biomechanics of this gymnastic exercise (and others) see the bibliography to http://www.coachesinfo.com/index.php?option=com_content&id=182&Itemid=148.
3 The tactic adopted by the governing body to discourage gymnasts from attempting dangerous routines in competition is to give a very low degree of difficulty score to such moves so there is no incentive to include them. This is what happened with the Gienger release move on the asymmetric bars.

Chapter 24

1 C. McManus, *Right-Hand, Left-Hand*, Weidenfeld & Nicolson, London (2002).

Chapter 26

1 In 2005, golf, karate, rugby union, roller sports and squash were considered for inclusion. Only karate and squash were selected for final voting and neither received the two-thirds majority support needed for inclusion in future Games, although karate received more than half the votes. In 2009 golf and rugby union got the required support to be added to the Games from 2016 and 2020 respectively. There are a maximum number of sports allowed in the Olympic programme in any one Games.

2 There have been many attempts to measure the force exerted by boxers and martial arts practitioners using sensors attached to punchbags. Typical recordings for top-flight fighters give forces of more that 4,500N for boxers, and 6,800 for kicks in Taekwondo.
3 Your hand should be able to withstand 20,000N but don't try this at home.

Chapter 27

1 The load must be applied vertically downwards, at right angles to the lever. If the load is applied at an angle to the lever then it is the component of the force at right angles to the lever that must be multiplied by the distance to the fulcrum.

Chapter 29

1 The British representative Billy Clarke led the race early on but faded to finish twelfth in a highly respectable time of 3hr 16m 8s. Two years later he ran 2hr 51m 50s on a grass track and improved to 2hr 41m in an indoor race in Dublin in March 1911.

Chapter 30

1 Only men competed in that first Olympiad but the women arrived at the second Olympiad in 1900. The first race in the modern Olympics was a heat of the men's 100m, won by Francis Lane of the USA in 12.2s. He finished third in the final on 10 April, which teammate Thomas Burke won in 12s.

Chapter 31

1 I. Palcios-Huerta, *Rev. Econ. Stud.* 70, 395–415 (2003) and P.-A. Chappori, S. Levitt and T. Groseclose, *Am. Econ. Rev.* 92, 1138 (2002); J. Haigh, *IMA J. of Management Mathematics* 20, 97 (2009).
2 A study of penalties in the German league over a sixteen-year period found that 76% were scored, although only 70% were scored in the 2005–6 Premiership season.

Chapter 32

1 Sometimes best of five will be played.
2 J. Haigh, *IMA J. of Management Mathematics* 20, 97 (2009).
3 This is because the marginal advantage for winning a set of m games

with marginal advantage $s^\star$ is $2s^\star\sqrt{(m/\pi)}$ but $s^\star = 2s\sqrt{(n/\pi)}$ is the advantage inherited from playing n points in each game.

4 The next alternative up would have been to play best of $n = 9$ but this would have required $m = 5$ (best of nine), say, to give $m \times n = 45$ and the games would end at 5 points – too short to be appealing. The next alternative down would have been $m = 3$ (best of five) with $n = 15$ for $m \times n = 45$: again, this is satisfactory but not enough of a change from the old system although it is used in national and local tournament play.

Chapter 33

1 See http://www.eracewalk.com/Tech.htm. Still frames can be viewed at http://www.racewalk.com/JPerez/JPerez.asp.
2 E. A. Trowbridge, Walking or Running?, *Math. Spectrum* 15(3), 77–81 (1983).

Chapter 34

1 Pick each of the odds by the formula $a_i = i(i + 2)$ to 1 and you can get $Q = 3/4$ and a healthy 30% return even when N is infinite!

Chapter 36

1 For full details see J. D. Barrow, Rowing and the Same Sum Problem Have Their Moments, *American J. Phys.* 78, 728–32 (2010).
2 http://www.rowhistory.net/topic.asp?TOPIC_ID=116 quotation from a *Rowing News* article by A. Anderson. Other stories exist about the origin of the Moto Guzzi rig. One claims that the crew rigged the boat wrongly (probably deliberately) and their coach, suspecting that they were trying to avoid rowing a planned time trial, told them they would have to do it in the mis-rigged boat and then do it again if they failed to make the target time. To everyone's surprise they surpassed both the target time and their own best performance over the course using their new rig.
3 Notice that because of the cancellation of the s and the fact that r always just ends up multiplying F, the problem is solved by finding all the ways to combine the numbers from 1 to 8 with four plus signs and four minus signs so the total is zero; for example, the four solutions I found correspond to the four ways to do this, which are $1 + 2 - 3 - 4 - 5 - 6 + 7 + 8$ for (a), $1 - 2 + 3 - 4 - 5 + 6 - 7 + 8$ for (b), $1 - 2 - 3 + 4 + 5 - 6 - 7 + 8$ for (c), and $1 - 2 - 3 + 4$

– 5 + 6 + 7 – 8 for (d). It can be shown that zero-wiggle rigs are only possible if the number of rowers is exactly divisible by four.

4 See V. Kleshnev, *Rowing Biomechanics Newsletter* 9 no. 104 (Nov. 2009), online at http://www.biorow.com/RBN_en_2009_files/2009RowBiomNews11.pdf.

5 There is a picture at www.life.com/image/82387609.

Chapter 39

1 The winner must score at least 15 points and lead by at least 2 clear points.

2 The winner must score at least 11 points and lead by at least 2 clear points.

3 J. Haigh, *Taking Chances*, OUP, Oxford (1999).

4 We want $R < RS + R2(1 - S) + RS(1 - R)$ which reduces to $1 - 2S < R(1 - 2S)$, i.e. $(1 - 2S)(1 - R) < 0$. But R is a probability and so cannot be bigger than 1 hence $1 - R$ cannot be negative and so we need $1 - 2S$ to be negative, hence $S > 1/2$.

5 The condition $p/(1 - p + p^2) > ½$ is that $(3p - 1 - p^2)/(1 - p + p^2) > 0$. Since p lies between 0 and 1, $1 - p + p^2$ is always positive. The condition that $3p - 1 - p^2$ is positive for p between 0 and 1 is $p > ½ (3 - \sqrt{5})$.

Chapter 43

1 The breaststroke, freestyle and backstroke styles were first distinguished by separate events at the 1904 Olympics in St Louis. The butterfly style, with the dolphin leg kick, emerged in 1935–6 as an attempt to create a faster version of the breaststroke. The dolphin kick violated the rules but the 1936 Berlin Games featured breaststroke swimmers using butterfly arm and breaststroke leg actions. Soon every breaststroker was doing this but butterfly was still not introduced as a separate event until the 1952 Olympics.

2 For a detailed hydrodynamical analysis of freestyle swimming see H. Toussaint and M. Truijens, *Animal Biology* 55, 17 (2005) which also reviews the early work of J. Counsilman and R. Schleihauf and collaborators in the 1980s which led to important changes in hand positioning and stroke motion that are credited with producing the dramatic reduction in swimming records during the period 1980–2000.

3 M. R. Kent, *Physics Education* 15, 275 (1980) and for more technical hydrodynamical analysis see R. E. Schleihauf et al., 'Propulsive techniques: front crawl stroke, butterfly, backstroke, and breaststroke', in

B. E. Ungerechts, K. Wilke and K. Reischle (eds.), *Swimming Science V*, Human Kinetic Bks, Champagne IL (1988), pp53–9.

Chapter 48

1 The force F needed for the hammer of mass m to move in a circle of radius r at speed v is $F = mv^2/r$. If the arm is kept straight and the hammer wire is fully taut then r stays constant and $F \propto v^2$. The range R of the throw after it is launched at speed v is proportional to v^2/g, where g is the acceleration due to gravity. Since F, the strength of a thrower of mass M and weight Mg, is expected to be proportional to $(Mg)^{2/3}$, we see that $R \propto v^2 \propto F \propto (Mg)^{2/3}$.

Chapter 49

1 http://www.youtube.com/watch?v=Thp YsN-4p7w.

Chapter 51

1 In the 1930s, times were only recorded to the nearest 0.1s and a record could only be broken by 0.1s so the maximum following tailwind for world record purposes was first defined to be +1m/s by the IAAF so that wind could not create a record-breaking performance. However, at their Congress in 1936, they changed the allowable wind limit to 2m/s and it has stayed the same ever since, even though race times are now recorded to an accuracy of 0.01s.

2 The drag force due to running through air at speed V with a following wind of speed U relative to the ground is proportional to $(V-U)^2$ whereas when running into a headwind of speed U the drag experienced is proportional to $(V+U)^2$. So we see that if $V = 10$ and $U = 2$ that the resistance with the tailwind is proportional to 64 but with the headwind is to 144, hence the asymmetry.

3 N. P. Linthorne, 'Accuracy of wind measurements in athletics', in A. J. Subic and S. J. Haake (eds.), *The Engineering of Sport: Research Development and Innovation*, Blackwell Science, Oxford (2000), pp451–8.

Chapter 52

1 Noting that $\cos(90° - B - C) = \sin(B + C)$.

2 We have also used the formula that $\sin(B + C) = \sin B \cos C + \sin C \cos B$.

3 Note $\tan 45° = 1$ and $\tan 30° = \frac{1}{2}\sqrt{3} = 0.87$.

Chapter 54

1 B. Berendonk, *Doping Documents: From Scientific Research to Cheating*, Springer-Verlag, Berlin (1991).

Chapter 55

1 By symmetry you are just as likely to be on the left of the central line to the goal as on the right. What we are calculating is the likely distance from it on either side.

Chapter 56

1 Rugby was last in the Olympics in the 1924 Paris Games but appears to have been discontinued after the USA beat France 17–3 in the final. The match was rough, and there were several injuries and serious crowd trouble. Police intervention was needed to escort the American team safely off the field.

Chapter 57

1 IPC rule 159.
2 I am most grateful to Franz Fuss for this and much other useful information about racing wheelchair design.
3 Amputees are excluded from this estimate. They introduce much greater variations and other stability factors.
4 Some materials have an added velocity dependence for the rolling friction but its effect is very small (about 4% of the μMg term) for the wheelchair so we neglect it.
5 F. Fuss, *Sports Eng.* 12, 41–53 (2009).

Chapter 58

1 There had been analogous events in France in the 1920s and 30s but with the order cycle–run–swim.
2 Another possibility is to calculate the mechanical work done on each stage and make this equal. This is harder to evaluate because of the role played by technique in the swim. You can use more effort to go slower with a bad technique. You would need to use an average performance as a standard for each event.

Chapter 62

1 The volume of the single cylinder in the top half is $\frac{1}{2}h \times \pi R^2$ and

the volume of the two half-cylinder 'legs' are $2 \times \pi(R/2)^2 \times h/2$, so the total volume = $\frac{3}{4}\pi hR^2$

2 The area of the top circle is πR^2 and the area of the outside of the cylinder of radius R and height h/2 is πRh, while the total area of the outsides of the two cylinders of radius ½R and height ½h is πRh, giving a total cooling surface area of $\pi R^2 + \pi Rh + \pi Rh = \pi R(R + 2h)$.

Chapter 65

1 For a spherical planet or moon of mass M and radius R, the acceleration due to gravity at its surface is $g = GM/R^2$, where G is Newton's gravitation constant.

2 When you are at a height h above the average radius, R, of the earth the change in g will be $g(h) = g(h = 0)(1 - 2h/R)$ so long as h is much smaller than R = 6,371km. Thus g also gets smaller as we go to higher altitude and the earth's gravitational pull weakens slightly. This gives a shift in g of about $2 \times 10^{-6} \times h$ when you go up to an altitude of h metres. This figure also allows for the extra mass of mountain that you go up to get to altitude h. If you just ascend to a height h in the air then the fall in g is slightly bigger, about $3 \times 10^{-6} \times h$.

3 If we calculate the effect precisely then the acceleration due to gravity g(L) at sea level will vary with the angle of latitude on the earth's surface as $g(L) = 9.780327(1 + 0053024\sin 2L - 0.0000058\sin^2 2L)$ m/s^2.

4 At the equator we have $g = 9.87$m/s^2 whereas at the poles $g = 9.832$m/s^2.

Chapter 68

1 C. Frohlich, *Am. J. Phys.* 49, 1125 (1981).

2 This was argued for in 1932 by J. A. Taylor, *Athletic J.* 12, 9 (1932) but was ignored.

Chapter 69

1 This was a predecessor of the present second string Europa League competition.

Chapter 70

1 The random data is taken from J. Wesson, *The Science of Soccer*, Inst. of Physics, Bristol (2002), p105.

2 The average performance expected is to obtain $(38 \times 3/8 \times 3) + (38 \times \frac{1}{4} \times 1) + (38 \times 3/8 \times 0) = 42.75 + 9.5 = 52.25$ points.

Chapter 71

1 It reflects the proportion of the kinetic energy of the air intercepted by the body that is destroyed.
2 If a following wind exceeds +2m/s the performance is not valid for record purposes.

Chapter 74

1 The period of the oscillation is $2\pi/w$, where w is the frequency of $\cos\{wt\}$.
2 R. B. Banks, *Towing Icebergs, Falling Dominoes and Other Adventures in Applied Mathematics*, Princeton Univ. Press, Princeton NJ (1998), p140.

Chapter 78

1 The component of the force of gravity down the direction of the slope is mgsinA where A is the angle between the slope and the horizontal, and sinA = vertical distance/distance up the slope = 1/G. This is the definition of the gradient, G.
2 Note that the horizontal distance is ½D and the distance up or down the incline is ½D × cosA where sinA = 1/G.
3 You should find $Thill/TL = (\frac{1}{2}\cos A)(V_L/V_d)^{1/3}(1 + V_d^3/V_L^3)$.

Chapter 79

1 This probability has been calculated in many places, for example J. Haigh, *Taking Chances*, OUP, Oxford (1999), p156.
2 Expand p^4 and $(1-p)^4$ in series keeping terms up to u^2. You will find that $S = (8u + 40u^2)/16u = \frac{1}{2} + 2.5u$ if terms in u^3 and u^4 are ignored because they are so much smaller than u and u^2.
3 Again, ignoring tiebreaks, the chance of winning a three-set match is $S_3 = S^2 + 2S^2(1 - S)$ and the chance of winning a five-set match is $S_5 = S^3 + 3(1 - S)S^3 + 6S^3(1 - S)^2$. See also G. Fischer, *Am. J. Phys.* 48, 14 (1980).
4 D. Jackson and K. Mosurski, in J. Albert et al., (eds.) *Anthology of Statistics in Sports*, SIAM, Philadelphia (2005), chap. 42.

Chapter 81

1 H. M. Toussaint et al., 'Biomechanics of swimming', in W. E. Garrett and D. T. Kirkendall (eds.), *Exercise and Sport Science*, Lippincott, Williams and Wilkins, Philadelphia (2000), pp639–60.

2 For world-class swimming speeds, the flow near the body will often become turbulent. The Reynolds number, $Re = vL\rho/\mu$, for a swimmer of length L = 2m with arms extended moving through water of density $\rho = 1{,}000\text{kg/m}^3$ and viscosity $\mu = 0.9 \times 10^{-3}\ \text{Ns/m}^2$ at average speed v = 2m/s is about 450,000 and the critical value for the sudden development of turbulent flow is about 500,000. The situation is therefore very finely balanced and slight increases in the speed about the average for the stroke cycle or variations around the swimmer's body will produce some turbulence.

3 Waves of length L are generated at a surface swimming speed v where $L = 2\pi v^2/g$, where $g = 9.8\text{m/s}^2$ is the acceleration due to gravity. This equals 2m for a speed of about 1.8m/s.

4 The rules allow a maximum of 15m underwater from the start and from each turn.

5 There is a video of his 'record' swim at http://www.wired.com/playbook/2011/02/dolphin-man-denied-record/?utm_source=feedburner&utm_medium=feed&utm_campaign=Feed%3A+wir ed%2Findex+%28Wired%3A+Index+3+%28Top+Stories+2%29%29&utm_content=Googl e+Feedfetcher.

6 H. Toussaint and M. Truijens, *Animal Biology* 55, 17 (2005).

Chapter 82

1 They were an amateur club (all Olympic competitors were amateurs then) who played in the first FA Cup competition in the 1871–2 season. They went out of existence in 1911.

2 They withdrew in 1924 and 1928 over disputes about professionalism. They were also unwilling to play matches against wartime enemies and this is why they did not join FIFA after World War I.

Chapter 83

1 L. J. Peter and R. Hull, *The Peter Principle: Why things always go wrong*, Morrow, New York (1969).

2 A. Pluchino et al., The Peter Principle Revisited, *Physica A* 389, 467 (2009).

Chapter 84

1 G. P. Brueggeman et al., *Sports Technol.* 4–5, 220–7 (2009).

2 P. G. Weyand, M. W. Bundle, C. P. McGowan, A. Grabowski, M. B. Brown, R. Kram and H. Herr, *J. Appl. Physiol.* 107, 903 (2009).

3 P. G. Weyand and M. W. Bundle, Point: Artificial limbs do make artificially fast running speeds possible, *J. Appl. Physiol.* 108, 1011–12 (2010). R. Kram, A. M. Grabowski, C. P. McGowan, M. B. Brown and H. Herr, Counterpoint: Artificial legs do not make fast running speeds possible, *J. Appl. Physiol.* 108, 1014–15 (2010).

Chapter 87

1 The exact solution for the downward speed at time t if the skydiver starts with zero speed at $t = 0$ is $v = U\tanh(gt/U)$ which approaches U as t becomes large and the distance fallen after time t is $(U^2/g) \ln[\cosh(gt/U)]$.
2 Kittinger used a small drogue chute to aid stability and control in the free-fall phase. The altitude record for a drogue-free descent is held by the Soviet Eugene Andreev, who descended without one from 25,460m in 1962.
3 The acceleration experienced in circular motion of radius r at angular velocity w is rw^2. A spin of 120rpm is an angular velocity of $120 \times 2\pi/60=4\pi$ radians per s. With a radius of 1.4m this gives an acceleration of $1.4 \times 16\pi^2$ $m/s^2 = 22.5g$, with $g = 9.8m/s^2$.

Chapter 88

1 W. G. Pritchard, Mathematical Models of Running, *SIAM Review* 35, 359 (1993).
2 The effects of temperature and pressure differences which feed into changes in the air density are much smaller than the wind effects.

Chapter 89

1 Arrows have a compound structure with hard and soft wood sections, the arrow head, and fletching, which all have different densities.
2 A fletcher (now a familiar English name) made arrows by hand and derives from the French *flèche*, for arrow.

Chapter 91

1 For simplicity we assume that the stops are sudden and ignore decelerations and accelerations.
2 This is the geometrical area times a drag factor, c, that reflects how smooth and streamlined it is. For a typical car, the geometrical cross-sectional area will be about $1.5 \times 2 = 3m^2$ but the effective area will

be about one third of this, giving $1m^2$. For a runner it will be about $0.45\ m^2$ and for a cyclist about $0.75m^2$.

Chapter 92

1 'Scuba' is an acronym for the 'self-contained underwater breathing apparatus' that was invented during World War II, and has become a word in its own right.
2 It seems to have been first proposed by Henry Power in 1661.
3 This is 14.7lb per square inch or 101.3kPa.

Chapter 93

1 The yield strength will be better than 50,000psi.
2 If you just stand still on the end of the board so that it is bent away from the horizontal by a distance x then the force kx is exactly balanced by your weight, mg, where m is your mass and $g = 9.8m/s^2$ is the acceleration due to gravity. A typical deflection of 0.75m created by a 65kg diver would give a value of $k = 65 \times 9.8/0.75 = 848$N per metre.

Chapter 94

1 This type of analysis of coin tossing was first carried out by Joseph Keller, *Amer. Math. Monthly* 93, 191 (1986) and then extended in greater detail by P. Diaconis, S. Holmes and R. Montgomery, *SIAM Review* 49, 211 (2007).
2 In general, if n is a whole number, 1, 2, 3, 4, . . ., then if N lies between 2n and 2n + 1 the coin will land on the hand with the initial face upwards and if n lies between 2n + 1 and 2n + 2 the initial face will land downwards.

Chapter 95

1 However, the Olympic marathon doesn't seem to attract all the top marathon runners. They often prefer to run in big-city marathons like London and New York. There are various reasons. Olympic Games can be held in venues with a climate that is unsuitable for distance running. Marathon runners can only compete once or twice a year at the marathon distance. If they have to run a qualifying race as well as the Games itself then they have no more chances to earn prize money in the lucrative big-city marathons. There are also no pace makers at the Olympics.

Chapter 96

1 You are allowed to gain torque for somersaults from take-off but not for twists.

2 This quantity is proportional to the mass times the angular velocity times the square of the radius of the circle in which the rotation occurs.

Chapter 97

1 N. De Mestre, *The Mathematics of Projectiles in Sport*, CUP, Cambridge (1990), p137.

Chapter 98

1 http://engineeringsport.co.uk/2011/03/01/the-heat-is-on-for-cyclists-in-the-london-velodrome/#more-1495.

Chapter 99

1 The coefficient of restitution, e, is the ratio of the speed before impact to that after impact when it is dropped to the ground from a height. The height reached after bounce is proportional to the square of the initial velocity after impact, ignoring air resistance. The Superball has $e = 0.9$ approximately.

2 R. L. Garwin, *Am. J. Phys.* 37, 88 (1969).

3 Equate the total energy of motion $= \frac{1}{2}MV^2 + \frac{1}{2}I\omega^2$ before and after a bounce, and also the angular momentum about the contact $I\omega - MRV$ before and after the bounce. Here $I = 2/5MR^2$ is the moment of inertia of the ball, ω is its angular velocity and V is the velocity of its centre. The component of the velocity perpendicular to the ground is reversed at each impact with $e = 1$.

4 For a very detailed analysis of Superball bounces, taking into account $e < 1$ for the bounces, see P. J. Aston and R. Shail, *Dynamical Systems* 22, 291 (2007).

5 This way of understanding the Superball motion is due to F. S. Crawford, *Am. J. Phys.* 50, 856 (1982).

6 A central high density concentration that creates a moment of inertia $1/3Mr^2$ rather than the usual $2/5Mr^2$ for a uniform sphere of mass M and radius r that gives the behaviour in the first figure in this chapter.

7 Children's snooker tables have this shape. Full-size tables are made from two adjoining squares.

Chapter 100

1 There is a detailed article about it by Steven Brams and Marc Kilgour, The Box Problem: To switch or not to switch, *Math. Mag.* 68, 27 (1995).

www.vintage-books.co.uk